CHRISTMAS TREES

Growing and Selling Trees, Wreaths, and Greens

by
Lewis Hill

A Garden Way Publishing Book

STOREY

Storey Communications, Inc.
Schoolhouse Road
Pownal, Vermont 05261

The information in this book is true and complete to the best of our knowledge. All recommendations are made without guarantee on the part of the author or Storey Communications, Inc. The author and publisher disclaim any liability incurred with the use of this information.

Cover design by Wanda Harper
Text design and production by Wanda Harper
Illustrations by Judy Eliason
Edited by Sarah May Clarkson and Gwen W. Steege

Printed in the United States by Arcata Graphics
First Printing, December 1989

Library of Congress Cataloging-in-Publication Data

Hill, Lewis, 1924-
    Christmas trees : growing and selling trees, wreaths, & greens / by Lewis Hill

    Bibliography: p. 145
    Includes index.
    ISBN 0-88266-566-9 (pbk.)
    1. Christmas tree growing—United States. 2. Christmas trees—United States—Market
ing. 3. Wreaths—United States—Marketing. 4. Christmas greens—United States—Market
ing. I. Title.
SB428.34.U6H55   1989   89-45222
634.9'75—dc20   CIP

Contents

Terms

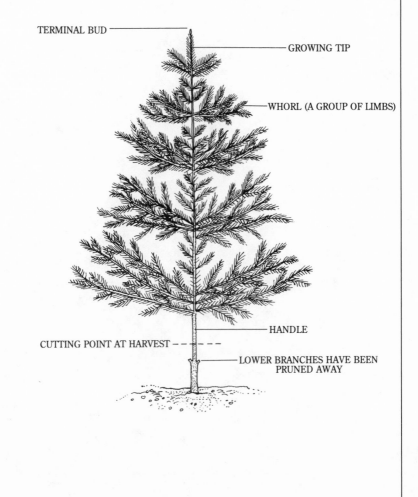

TERMINAL BUD ————————————

———— GROWING TIP

———— WHORL (A GROUP OF LIMBS)

———— HANDLE

CUTTING POINT AT HARVEST – – – – –

———— LOWER BRANCHES HAVE BEEN
PRUNED AWAY

I

GETTING STARTED WITH CHRISTMAS TREES

Christmas trees and greenery have become such an integral part of modern life that it is hard to imagine a December without them. Few streets in any city, village, or rural area are without homes decorated for the holidays, brightening up what would otherwise be the gloomiest, most depressing time of the year. Twinkling lights strung around apple trees and window frames are attractive, but no lights lift the spirits quite like those on an honest-to-goodness, fragrant evergreen. Furnishing the many trees used for homes, businesses, and public buildings has become a multimillion dollar business. In this book, we will guide you through all of the steps necessary to maintain a successful Christmas tree operation, from selecting your site and the species to grow on it, through caring for your plantation, to harvesting your crop and managing your business.

CHAPTER ONE

Money Can Grow on Trees

T he many Christmas trees and greens that decorate homes and businesses everywhere come both from large Christmas tree plantations of thousands of acres, as well as from small farms like our own. Small growers either market trees locally or sell to wholesale buyers who make up tractor-trailer loads by buying from several producers. The trees harvested each year by large and small growers range from 2-foot tabletop miniatures to giants such as the one cut each year for Rockefeller Plaza in New York City.

Some small-scale tree farmers plant only a few hundred trees each year either as a boost to their regular income or as an investment and pleasant diversion in retirement. Other growers plant all their acreage at once, so they can harvest two or three big crops as a child reaches college age or for a once-in-a-lifetime trip abroad. Rural youngsters often raise trees as a conservation project for Scouts, 4-H, or Future Farmers of America; and civic organizations buy or lease land to plant them as a money raiser for community projects. Some people raise a few trees just for the fun of it — for their own use and to give to family and friends.

Not only can a tree farm help you make money with interesting outdoor work, but it makes excellent use of fallow land. A field of Christmas trees is attractive and it doesn't block views as a forest does. Furthermore, such a project is ecologically sound with each acre of trees providing daily oxygen for eighteen people, erosion control, and excellent cover for wildlife.

It is often hard, even on paper, to figure out how to make agriculture projects profitable. Chicken farmers must be extremely efficient to turn a profit, as must anyone who raises turkeys or potatoes. But how can one go wrong with a crop of 1,200 inexpensive trees on an acre, for which you wait six to ten years while they grow, then sell them for $10 each and pocket $12,000? If you had 20 acres and planted only 2 acres each year, you could earn $24,000 a year, forever. Not bad for a spare-time activity!

It is possible for things to work out in such a rosy way, but to win that jackpot you must not make any mistakes, and the weather, markets, and prices must be all favorable. There are other potential clouds on the horizon, too. Each year far more seedlings are planted than could ever be sold as Christmas trees, and artificial trees continue to cut deeply into the market for real trees. No one can be sure what the situation will be 10 years from now when this year's seedlings are ready for harvesting.

Even with the risks, one thing that weighs heavily in the grower's favor is the nearly recession-proof market. People tend to spend money on a Christmas tree, even if they have to give up an evening out or some other luxury. Based on past experience, most of us growers are confident that the demand for fresh, real trees will continue, and that we will be needed to fill it.

Small-Scale Production

Compared with the huge plantations growing millions of trees and employing hundreds of workers, our tree lot is relatively tiny. Over a period of six years, beginning twenty-two years ago, we planted 24,000 trees on about 20 acres. Although we waited about twelve years for the first harvest, we have been cutting and replanting now for over ten years. We did not shear or harvest about 2,000 of the original planting, but use them instead for growing greenery that is clipped each fall and made into wreaths and other

The Traditions of Christmas Trees

The tradition of decorating a holiday tree got its start in pre-Christian times, many centuries ago when Northern European tribes held festivals and religious rituals during the dark wintry days to induce the return of the sun. Evergreens — symbols of life because they don't die over the winter — were brought into homes and places of worship as part of these ceremonies. In the sixteenth century, Martin Luther is said to have decorated a fir tree with candles to represent the starry heavens from which Christ had come. As the custom spread, others added apples to represent the Garden of Eden, and cookies as a symbol of the Eucharistic wafer.

Although the Pilgrims and Puritans regarded any celebration of Christmas as pagan, other European immigrants to North America made the holiday a joyous occasion, and decorated with fir trees and holly. Hessian soldiers fighting for the British in the Revolutionary War probably introduced the lighted tree in the 1700s, and in 1856 President Franklin Pierce set one up in the White House. Not until 1923, however, did Calvin Coolidge light the first National Christmas Tree on the White House lawn. Now this impressive ceremony each December is only one of many taking place at state capitols, city parks, and little village greens all over the country.

holiday decorations.

Many evergreen plantings are much smaller than ours. Some consist of only a few dozen trees of various ages growing in the back corner of a suburban or village lot where they form part of the landscape. Each year a few of the trees are cut, either to be sold, used in the home, or given away, and new little trees planted to replace them. Plantings of this size can be cared for with ordinary garden equipment.

Small Investment Required

One very favorable aspect of a small-to medium-sized Christmas tree project is the surprisingly small cash outlay needed for planting, annual care, and harvesting when compared with that of many other rural enterprises such as raising livestock, operating a dairy farm, growing vegetables, or starting a plant nursery. The planting stock is inexpensive and the few essential tools needed for a small plantation can be used for many other home and garden projects. Although you should not underestimate the amount of work needed to grow premium trees, highly skilled labor is not necessary, and much of it can be done in your spare time or by students and others willing to moonlight during evenings or weekends.

The wait for the first profits can seem depressingly long, but the time is actually little more than it would be for the first big harvest of apples, walnuts, blueberries, or other crops, and it is far less than waiting for a planting of firewood, timber, or syrup-producing maple trees to mature.

Annual Harvests Versus a One-Time Bonanza

Unless you have a good reason for wanting the entire income from a Christmas tree project within two or three years, stagger your plantings over several years. One mistake beginners in the tree business are prone to make is to plant all the available acreage at once. There are things in favor of doing this, of course. If you have 20 acres, it is certainly more efficient to hire a planting machine and finish the whole job within a few days. Twenty-four thousand trees all the same size look pretty neat, too.

Unfortunately, trees planted at the same time need the same amount of shearing and other care each year, and most will become salable the same year. This schedule could create difficulty in finding a market for a large number of trees, especially when you have had no previous experience in dealing with buyers. Your one and only crop could also be ready to harvest during a period when there is a surplus of trees that would make it difficult to sell them at any price. Even if you are in luck and they can all be sold at a good price, you will be faced with a huge harvesting job that must be completed within a few short weeks, and you will have to pay big income taxes that year. Then you will be left with no more trees to sell until the next crop matures perhaps ten years later.

Planting a portion of your acreage each year not only provides a steadier annual income, but also more evenly distributes the work of planting, cutting, shearing, harvesting, and marketing. Equally important, it puts you in a better position to deal with buyers who want to be certain of a source of trees they can depend on year after year.

Alternatives To "Do It Yourself"

If you want to put idle acres to work, or like to see something growing on your property but have no time to take on such a project, consider hiring a local professional to take over all or part of your growing enterprise. Absentee landowners, especially, find this a good way to manage a plantation. In most areas where trees are grown, there are contractors who do planting, shearing, harvesting, and possibly even the marketing; or a neighboring tree farmer may be willing to operate your plantation along with his own. Not only will you save the cost and headaches of buying and maintaining equipment, but you can also avoid the bookkeeping and tax problems that come with paying laborers.

Unless you are skilled in legalese, have a good lawyer draw up the contract between you and your contractor to clearly set out the time frame of your agreement. When one of our neighbors decided to sell his farm, he found that a contract to cut a crop of timber on the property nearly a hundred years previous had never been canceled, and this was considered a "cloud" on his deed. He couldn't locate any of the heirs of the timber buyer and had considerable trouble satisfying the lawyers who were representing his buyer.

Working With Wild Trees

Not all Christmas trees are grown in straight, well-spaced rows. Wild tree farms are not as common as they once were, but they can still be found where evergreens grow naturally and reproduce readily without replanting. In our area, for example, balsam fir and white spruce trees grow so easily that dairy farmers must frequently cut or brush-hog them from their pastures in order to keep grazing land open. If you own or buy land that has an abundant supply of a salable variety of small evergreens already growing, it is possible, with some work, to begin selling them in only three or four years, rather than having to wait a decade for your first harvest. You can probably make money even the first fall by selling the greens salvaged from the pruning and thinning operations.

In addition to an early income, we have found there are other advantages in working with wild trees. For one, both the expense of buying seedlings and the time planting them is avoided. Trees that have seeded naturally grow faster, too, because they are already suited to the soil and climatic conditions and don't suffer the trauma of being dug and replanted.

I don't want to mislead you, though. A wild stand involves a great deal of labor in removing species you don't want, thinning out the remaining good seedlings, and shearing them skillfully to compensate for the

Current Data on the U.S. Christmas Tree Industry*

- 85 to 95 million new trees are planted each year.
- 1 million new acres are planted each year.
- 100,000 people are employed by the Christmas tree industry.
- State and federal taxes each year indicate the Christmas tree industry produces a crop with a wholesale value of $322,000,000 and a retail value of $695,000,000.

Source: National Christmas Tree Association, Inc., 611 East Wells Street, Milwaukee, Wisconsin, 53202-3891
*1987

poor growth caused by their previously crowded conditions. See pages 63-66 for more information on how to handle this type of tree farm.

In a world where so many jobs are dull and uninteresting, we feel lucky to be doing something we really enjoy — growing trees. It is rewarding to work outside with nature to create products that look so nice when they are growing and bring happiness to so many people after they are harvested. I'm not sure we would continue if money didn't also grow on those trees, but it would be tempting.

CHAPTER TWO

The Burgeoning Christmas Tree Industry

In the thirties and forties, when I was growing up, the dairy farmers in our neighborhood had begun to sell wild balsam fir and spruce trees from their pastures. Most made a deal with an itinerant buyer who paid a few cents per tree, cut what he needed, then hired the farmer to haul them to the railroad to be transported to urban areas. A few farmers, concerned about dishonest buyers and the messy job they did of cutting, began to harvest their own. Some enterprising souls even hunted up city markets and shipped trees directly to them.

The farmers were not too concerned about the quality of the trees, and anything that was green and shaped somewhat like a tree was tied into bundles of three to six and shipped, although they would not have dreamed of having such shameful specimens in their own living rooms. No one considered thinning the wild trees so they would have room to grow, nor shearing them for a tighter form, nor fertilizing for better color. For many years a tree had been considered something to hide behind or a weed to dispose of so there would be more grass to feed the cows and horses. Establishing a tree farm would have been unthinkable.

It was the "folks from away," as we called newcomers to our area, who showed us how scrubby pasture trees could become as profitable a crop as milk or potatoes. They bought abandoned, worn-out farms at low prices, planted trees in neat, straight rows and, much to everyone's surprise, appeared willing to wait ten years for their first income. Puzzled dairy farmers shook their heads when they observed tree growers buying fertilizer, weed killers, and insect sprays, and then wasting valuable summer days shearing their little trees into bushy specimens. But these same farmers who were accustomed to spending an entire day each December searching through the woods for a Christmas tree that would please their family, were dumbfounded a few years later as they drove by acres of the newcomers' bright green, perfectly shaped firs, spruces, and pines.

Not everything turned out to be apple pie and maple syrup for the growers, however, and dreams of easy money vanished for some when they found they could not sell their acres of Scotch pine because the trees had developed crooked stems or needles that turned a sickly yellow shade just about the time they were ready to be cut. Others, who tried to make use of land that was unsuitable for other crops, discovered that the species they were growing did not do well in sandy gravel or swampy soil. We were among those early growers who were completely unprepared for the ravages of deer, mice, weather, disease, and insects.

As newcomers to tree farming we found it quite different from raising a garden or other crops. Anyone accustomed to putting a corn seed or tobacco plant in the ground in the spring and watching it grow to 8 feet in three months had to adjust to giving trees a decade of care before they reached the same height and could be sold. Those who had formerly grown nursery stock, as we did, discovered that a Christmas tree must be raised to look its best in December, an unusual time to sell outdoor plants.

Selling the produce was often an unpleasant experience, too. Many early growers produced excellent trees only to be victimized by dishonest buyers who gave them bad checks or unfulfilled promises. "We'll pay when the last load is picked up," they would say; but they never appeared for that load. As in every new business, experience proved the best teacher. Foresters, lacking training in Christmas tree growing, at first were reluctant to recognize the fledgling industry, but they eventually became excited by the new opportunities and helped to disseminate information and form growers' organizations. Within a few decades wholesale prices had jumped from 5 and 10 cents for a wild pasture tree to $20 or more for each exceptional plantation-grown specimen. The undecorated wreaths we sold in the 1950s for a quarter, now command more than sixteen times that.

Future Prospects

Even the experts disagree about the potential for the prosperity of the Christmas tree industry. Pessimists predict that a surplus of trees is inevitable, since an estimated 85 to 95 million trees are planted each year and only about a third of that number are likely to be sold during the holiday season. Others disagree, noting that every year for the past thirty-five, far more trees have been planted than could possibly be marketed, yet growers have continued to sell their trees. They point out that, as a rule, fewer than half the trees planted are given enough care to become salable. Some die at an early age because they were planted poorly to begin with, dried out in a drought, or were the wrong species for the climate or soil. Some are damaged by rodents or larger animals, or suffer from diseases, insects, or snow and ice damage. Of the trees that survive, many grow into bad shapes because they haven't been sheared or pruned properly, or were choked by weeds, grass, or brush. Would-be growers often become tired of waiting for the harvest or they simply lose interest and let their plantations grow into a forest.

According to the doomsayers, that was yesterday's scenario. Today's growers, both in the United States and Canada, are more professional, they say, and more knowledgeable than those of past years, so it is likely that a much higher percentage of their seedlings will grow into premium trees. Therefore, a serious marketing problem is likely to develop when these millions of trees reach harvestable size.

Pessimists also predict that the market, instead of growing as the population increases, will actually shrink, the villain being the artificial tree. The first plastic trees looked phony and were popular only in shopping centers, stores, and offices, where a tree would be on display for several weeks. The new models have become so realistic-looking, however, that they are now seen in many homes. Ordinances forbid natural trees in some buildings because of the potential fire hazard and this trend is likely to accelerate. Growers fear, too, that a new generation has arrived with no tradition of enjoying a live tree and little concern for whether a tree is fake or real, especially if selecting a real one means a cold trip to an outdoor lot on a stormy, wintry day. Then there is the price. Real trees have

become an expensive luxury in recent years, and even though the fakes are more costly, shrewd customers wait until the after- holiday sales and buy one at half price for future holidays. They know that the artificial tree will last for years, be easy to store, drop no needles on the rug, need no water stand, and be fire-resistant.

For many years Christmas tree farmers refused to believe that the public would actually buy a phony tree if a high- quality natural one was available. Most, however, have now accepted the fact that the artificial tree is here to stay and a real threat to the industry, so they have begun to take action to increase the appreciation of a real tree. Christmas Tree Association leaders are promoting the advantages of natural trees, many strains of which have better color and better needle retention than ever before. They are also pointing out to an ecologically minded public that a live Christmas tree is a renewable resource, unlike the plastic models that use up already scarce petroleum reserves.

On the positive side, too, there is evidence that the public is tiring of a plastic, artificial world. Increasingly, people are more appreciative of the real thing and want to experience the sights and fragrances of former days. Optimists also think that more efficient cultural methods can result in lower prices, making real trees more attractive to buyers. Growers now plant transplants instead of seedlings (see page 39), interplant between older trees that will soon be cut, and use fertilizers and herbicides more wisely, reducing by one-third the growing time formerly required. The use of power shears, mechanical tree cutters, loaders, and other machines has eliminated much of the hand labor needed for production and harvesting. New strains of trees have better disease resistance and require less shearing, so growers save time and money in production. And if all else fails, growers tell themselves that if they can't sell their crop as Christmas trees, they can let it grow into timber. There is every evidence that a growing population is not likely to have a surplus of building materials in future years. When you weigh the pros and cons of growing any agricultural crop, you always find it a gamble. Working with the many whims of nature is risky and the whims of the buying public can be equally unpredictable. So, should you grow trees to make money? One grower told me that even though he had a chance to sell his farm to a developer, he had no intention of quitting. "I never play the stock market or the lottery," he said. "Christmas tree growing is an exciting enough gamble for me!"

CHAPTER THREE

Selecting the Best Site

I f you are planning to buy land specifically to grow Christmas trees, you have the advantage of searching for the perfect location — land that is accessible to a good road, with fertile soil suited for the species of trees you want to grow, and perhaps even a parcel that is blessed with an ideal climate.

You may also be able to choose property that is near the market you want to supply, because this factor can greatly determine the amount of money you can profitably invest. In locations far distant from city markets, the price per tree may be only one-half to one-third of what it might be near an affluent suburb.

Accessibility to a roadside is important because trucks should find it relatively easy to drive in off the road and turn around, if necessary, at the time they want to pick up the trees when buyers want them in November or December. Adequate drainage is essential. Good summer roads may turn to something like

quicksand when they become muddy or snowy in early winter. The harvest season is so short that if transportation problems arise, they will be costly both in time and money. The lay of the land and steepness of a slope are also important considerations when you're looking for a plantation site. Most Christmas tree species can be grown successfully on steep northern slopes that are not suitable for other crops. They can also be grown among rocks in spots that would be intolerable to other plants, so small pieces of otherwise useless land can often be used for this purpose. Such locations are usually impossible for machinery, however, which means that planting, mowing, fertilizing, and other care must be done by hand.

Late spring frosts can damage the early, new growth of trees, so avoid planting in frost pockets. Gentle northern slopes that warm up slowly in the spring and thus delay early growth are ideal, as are higher areas where frost can slip easily to a valley below. If you don't have such a location, plant only frost-resistant strains of trees (see pages 25-38). Usually the best choices are native species, because they are already acclimated to local conditions and seldom start to grow so early that they are damaged by frost.

To create a successful plantation, the selection of a good site must be made in tandem with your choice of a suitable species. Many species of trees grow well in a wide variety of soils and in different climates, but all do best when conditions are to their liking (see pages 25-38). If you are able to choose from a wide range of locations, use aerial, topographic, and soil maps available from your extension service to help you make a decision.

Evaluate your available sites with an eye to their surroundings, especially if you plan to use chemical fertilizers or pesticides. Be very cautious about planting trees on land near residences that use well or spring water, where development may occur later, or on land that slopes toward a river, reservoir, or other body of water. Chemicals move through the soil slowly, but over the years they can travel great distances, and in recent years nitrates, herbicides, and insecticides have polluted many water supplies. Even if you do not use any dangerous chemicals, lawsuits and criminal actions frequently arise even from suspected pollution. Unfortunately, in such cases the accused is not always presumed innocent until proven guilty, and any action can result in a lot of expense and unwanted publicity.

Checking out the Soil Depth, Fertility, and pH

Many advertisements for seedlings state that Christmas trees are an ideal crop for poor, worn-out soil, which implies that they will thrive where other crops will not. It is true that nature often covers neglected pastures with evergreens of one kind or another, but because of the lack of humus and nutrients, trees in such locations usually grow slowly and have poor color. For a successful tree planting, look for fertile, deep soil or, lacking that, soil that can easily be improved. On our own acreage, which had formerly been a dairy farm, we planted seedlings in two different locations: a series of small, rocky pastures with thin, worn-out soil that had begun to grow up to a mixture of evergreens and hardwoods, and a few open hayfields where the soil was rich and covered with a heavy sod of timothy and clover. Those in the fields grew beautifully, with excellent color, but the trees in the pastures grew much more slowly and had a decidedly yellowish tinge. The difference between the two plantings was so pronounced that many visitors were convinced we were growing two different species of trees and some even asked where we bought the premium planting stock in the fields.

Learn something about the former use of your land. If heavy applications of lime have been made over the years in order to grow clover or alfalfa, the soil may be too alkaline for pines and perhaps even other species (see pages 25-38). Take soil tests at different depths if you have questions.

If crops such as potatoes, corn, or nursery stock have been raised on the site, ask if any chemicals used there may have a harmful residual effect on your trees. In heavily grazed pasture land, dig down in several areas to assess the depth of soil and to determine whether surface water may collect and possibly drown tree roots when temporary floods occur during heavy rains or melting snows.

Take a good look at the condition of the soil on the plot of land you will be planting. Good soil contains a high amount of humus (partially decayed organic matter) that is able to absorb both moisture and fertilizer and release them later as the tree needs them. Humus also serves as a valuable buffer — soils well supplied with it are more forgiving of chemical excesses, and more tolerant of elevated acidity or alkalinity. Soil in good condition needs little or no improvement before planting. It can produce crops of trees with good color and heavy needle density indefinitely, with only an occasional feeding.

If you don't feel qualified to judge the condition of your land, ask your county agent, forester, or soil conservation official for advice, usually at no charge. Such experts may be able to spot conditions that are not readily obvious, and to be more certain, they may also recommend soil tests or foliar analysis.

Excessive amounts of sulfur and peat combined with a lack of calcium can cause the soil to be overly acid. Too much calcium can cause it to be excessively alkaline. Either condition can lock up soil nutrients, mak-

ing them unavailable during the growing season. Soil acidity/alkalinity is measured on a pH scale of 0 to 14, with 7 being neutral. Readings of less than 7 indicate acidic soil, more than 7, alkaline soil. Most soils fall between 4 and 7, although some peat bogs may be as low as 2. Crops such as alfalfa, clover, and blueberries that are growing in soil outside their preferred pH, will do poorly. Fortunately, Christmas trees are more tolerant and most species grow well in a wide range of soils as long as they contain sufficient amounts of humus and nutrients. Eastern white cedar or American arborvitae *(Thuja occidentalis)* is one of the few evergreens that thrive in less acid soil (pH 6.5 or higher), but they are not widely grown for Christmas trees. Even though other species are not as fussy, however, each has a pH preference. White spruce and balsam fir do best in soils of 6 to 6.5; red and Colorado spruce, 5.5 to 6; and most pine species prefer a more acidic soil, from 5 to 6.

By observing the plants already growing on the land, you can make an educated guess as to the soil acidity. Count it as a particularly good sign when a good number of the tree species you want to raise are already growing in the area. Where clover and goldenrod grow lavishly, the top few inches of soil probably has a pH of 6.5 or higher. An abundance of white cedars and "hard" water in nearby wells indicate a high pH at a deeper level. Sorrel, laurel, and wild blueberries are good signs that the topsoil is acidic; if you find oaks, pines, and hemlocks growing naturally (in other words, where someone didn't plant them), you can be sure that the subsoil is acidic also.

If you don't feel confident about deciphering the soil acidity by observing the plant life in the area, invest in a soil test kit, which is inexpensive and available at most garden supply stores. Test the soil in many different places and at different depths. Tree roots go deep, and if lime or wood ashes have been spread during recent years, the upper soil level will probably test quite differently from that at a depth of 12 to 24 inches.

Clearing Land for Tree Planting

Clearing land that is growing large trees, and building up the soil so that it will successfully support Christmas trees is usually not practical from an economic viewpoint. If, however, the soil on a wooded site looks like it could be improved, and the trees are mostly ones with trunks less than 4 inches in diameter at ground level, it may be worthwhile to convert it into a tree plantation. The ideal method to do this is to cut down the existing trees, bulldoze out the stumps, then till the land and grow a cover crop for a year before planting seedlings. It is not a good idea to plant new trees among the old stumps unless they are very small, and even then considerable fertilizer will be needed to get the trees to grow well. Much of the land where trees are now growing was probably once farmed and subsequently abandoned, so there are likely to be only a few inches of nutrient-deficient soil over a layer of heavy clay or gravelly subsoil.

Climate

If you already own the land where you plan to grow trees, you can't do much about climate. Your plantings should thus be limited to the species that will grow well at your location. If you are shopping around the country for the proper real estate, on the other hand, you can be more choosy. Wherever you plant, of course, there are likely to be occasional surprises in the weather.

Douglas fir thrives in western North America where it grows naturally, but it can be fussy about climate and soils in other regions. The various strains differ considerably, however, and seedlings from trees grown near the coast are less hardy than seedlings grown from those at higher elevations.

Most firs and spruces need cool summers and long, cold winters to thrive, so they grow well only in the northern states, southern Canada, and the higher elevations of the Appalachians. The pines grow well over a much larger area and some species thrive in southern and southwestern regions.

When searching for a good growing climate, look beyond the average highest and lowest temperature readings. Adequate rainfall is also very important for good tree growth, especially during the weeks in early summer when evergreens make their visible top growth, and through the late summer and fall as the roots continue to grow. This is when the trees store up the necessary nutrients and moisture that keep them green and fresh-looking after harvest. Christmas trees are irrigated on some farms, but this is expensive and requires a good supply of water.

Checking a climate zone map can be useful, but keep in mind that weather patterns often vary widely within each zone. Miniclimates can make frost-free seasons several weeks longer or shorter than they are only a few miles away. Elevation, wind, slope, nearby lakes or large rivers all help to determine whether you can expect early fall and late spring frosts, frequent hailstorms, high winds, or heavy snow and ice storms. If you are buying a new planting site, talk to nearby residents and check with the extension service before investing money. There's no way to anticipate once-in-a-lifetime hurricanes, tornados, and floods, but at least we can try to avoid normal adverse climatic conditions.

It may be well to check out the neighborhood, also. Neighbors with cattle, horses, goats, and sheep are not always careful about keeping them confined, and these can all be hard on Christmas trees. Cross-country skiers, or people who ride all-terrain vehicles or snowmobiles, may be a problem. Wild animals such as deer, moose, elk, and even some of the smaller creatures such as porcupines, can also be menaces, so you might want to check on their density and figure out the best ways of coping with them before planting. If a fence is going to be necessary to keep pests out, it will add considerably to your costs, both to build and maintain.

CHAPTER FOUR

Your Investment In
Time and Money

When we were starting in the Christmas tree and nursery business, I stopped to talk one day with a French Canadian who had recently bought a farm in our neighborhood. His English was still shaky and my French was worse, but we were both trying hard to understand each other. Finally, he said in disbelief, "You don't have no cows, all you do is grow trees? You like that better than working?"

Although by "working" he had meant dairying, his remark reminded me that a lot of people think there isn't any actual work involved in growing trees: "They just grow, don't they?" Even the firms that sell Christmas tree seedlings and transplants are not usually realistic about how much labor time and expense you will have to put forth before you get a wad of cold cash in your hot little fingers. No matter what the catalogs imply, buying and planting the small trees will not be your last or, in most cases, even your biggest, investment before you begin harvesting.

Before you order the first tree, it is important to plan carefully an investment budget of both time and money — you will probably find that the time involved is harder to estimate than the money. Small operations can be strictly a spare time occupation, but you must be willing to devote a few summer weekends to controlling weeds, insects, and disease, and to doing the necessary pruning and shearing. Studying the Grower's Calendar on pages 135-36 can give you an idea of the various chores that will be necessary at different times of the year.

It is difficult to explain what is meant by small, medium, or large operations. We think of a small planting as something that can be managed pretty much with the tools and equipment that a gardener or small farmer would ordinarily have on hand. One might decide to buy a tree tyer or wrapper, or one might sell the trees loose or even tie them by hand. Up to

A Christmas Tree Garden for Small-Scale Production

A few weeks ago, a retired doctor in our town called to ask about growing a few Christmas trees in his backyard. He wanted only one for himself and a few to give away each year, he said, and since he had plenty of land, thought it was foolish to spend days hunting for a good wild one or driving twenty miles to pay for one grown on a plantation.

Many others have had this same thought and are raising trees in their own backyard on a small scale. The only space necessary to grow 100 trees of various sizes is a plot 50 by 50 feet, a mere sixteenth of an acre — about the size of a small vegetable garden. By replanting new trees as you cut, you can harvest from seven to ten trees a year indefinitely. The cost is little more than the labor involved, with no sales tax to pay. If, in addition to trees for yourself and for friends, you want to harvest a few more trees per year, perhaps for extra income or as a community fund-raising project, by increasing the size to 1 acre you should be able to produce 100 to 150 trees a year. Naturally, the planting and managing of a small lot is less demanding than on a large plantation, and the start-up investment considerably less. Most of the work can be done with ordinary garden hand tools and the weeds controlled with a lawn mower.

When a tree-growing project is a small, part-time enterprise, it is easy to forget about if it is tucked away on a back lot where it is seldom seen. Locate it, if possible, where you will see it daily and notice when shearing, mowing, and pest control is necessary, and where you can also guard it from pillage.

Buying only a few seedlings or transplants can present a problem because most wholesale nurseries cater to large growers and won't sell fewer than a hundred of a single size and species. Some do supply smaller

(continued)

10 acres could be considered a small operation, harvesting about 1,000 trees on one acre per year.

A medium-sized operation, perhaps up to 50 acres, would probably require extra help, as well as more power equipment, some of which might be rented. A large planting is usually thought of as something in the hundreds of acres, involving a great deal of equipment, labor, and investment.

Our tree operation worked out well because it was a manageable size. We kept expenses low by using equipment we had on hand and doing the work ourselves. With several 4-H boys, we pulled 24,000 balsam fir seedlings from our woods over a period of several years and hand amounts, however, or you may be able to find an ornamental nursery that is willing to share some of the small trees they plant for growing on to landscape size.

Another way to get planting stock, if you have time and patience is to grow your own — a project that is satisfying and fun. Buy tree seeds in small packages from a seed company (see Appendix), or collect your own and plant them in flats or beds (see pages 43-45 for details).

A much faster method of obtaining your own stock is to look for native seedlings. They often grow wild in places where farmers are happy to get rid of them, so it is easy to get permission to pull a few each year. If you find a spot where you can dig a variety of sizes, you may be able to gather and plant at one time enough trees for a half dozen annual harvests. If a tree is over 15 inches tall, always dig and move it with a ball of earth and plant and water it immediately. Continue to water it heavily once a week for at least a month. (See pages 41-42 for further suggestions on how to do this.)

Unlike large producers, most small-scale growers like a variety of sizes and don't want all their trees to be the same size when they cut them. To get the sizes you want each year, you can adjust growth to a large degree by controlling the amount of fertilizer you use. If you want some trees to mature faster than others, feed them a bit more but be a little stingy with those you want to hold back for future Christmases. You can also keep trees small for an extra year or two by shearing them more severely than normal.

We use these techniques in our own plantings, and the trees we gave away last Christmas ranged in size from a 3-footer for a stand in a cousin's small apartment to a 15-foot specimen that graced the front of the local church.

planted them. Since we already owned the land, our first year's investment was in labor only, mostly on evenings and Saturdays. The second year, after the trees had become well established, we applied an herbicide to eliminate part of the grass and weed competition around each tree and, where the soil was poor, applied 10-10-10 fertilizer. After that we let the grass grow, and the shearing was done with the same hand tools we used to shape evergreens in our nursery and landscape business. Fortunately, it wasn't necessary to spray. We harvested the trees with the same chain saw we use to cut firewood and remove cull trees from our nursery. As our plantation ages and needs replanting, however, weeds and hardwood brush are becoming more of a problem, so more time and money will be needed to control them.

Larger Plantings

If you plan a plantation of more than a few acres, your investment in time and money will naturally be much more. Unless you hire a lot of labor, considerable machinery will be necessary, and you must also plan on the expense associated with maintaining and storing it. Since most equipment will be used only a few days a year, it may be wiser to take advantage of the services of professional tree contractors in your area for such jobs as planting, spraying, or shearing; or you may be able to rent equipment from them. Many contractors are tree growers themselves, and like to keep their employees and machinery productive during the times they aren't needed on their own plantation. There can be drawbacks to dependence on others, however. Contractors, to be sure of plenty of work, often promise more than they can deliver, and bad weather may upset their schedules even more. Planting and shearing, especially, must be done within critically short periods, and I have known growers who, frantic and frustrated from waiting for others, ended up by buying their own equipment.

Some of the things you will probably need on a large tree farm are a tree planter, power mower, power shearer, fertilizer spreader, sprayers, and tree harvesting equipment. There will also be the annual expenses of fuel, storage of equipment and supplies, fertilizer, machinery upkeep, pesticides, and labor — all of which can add up to a sizable amount before the first tree is sold.

Before you buy new equipment, you may want to investigate the newspapers and trade magazines for classified ads offering used equipment, and check out auctions of growers who are discontinuing operations. Christmas tree associations often have "buy, sell, or trade" classified ads in their newsletters. You may find worthwhile offerings or perhaps place advertisements for the items you need.

Early Christmas tree growers had little choice in power equipment, most of which was originally designed for some other purpose. Now there is a huge variety of machines for planting, mowing, shearing, and harvesting, and new ones are introduced each year. To make the best possible choice for your own needs, take time to study catalogs, visit agricultural fairs, and attend Christmas tree meetings to get up-to-date information on both what tools and machinery are available and what other growers a finding useful. Some suggested equipment is included in the chapters on planting, shearing, and harvesting.

When you buy power equipment, make sure that future servicing and parts will be available when you need them. An imported power saw with a cheap price tag may seem attractive, but if local servicing is not available, you can lose a lot of valuable time while you wait for parts. As every forester and farmer knows, breakdowns never occur except during those times when the device is most needed.

Tractors, trailers, and trucks are necessary on a large plantation, and can also be useful in a medium-sized operation. Many expense-conscious tree farmers have found that they can do much of their work with a 4-wheel drive pickup truck, and they hire a nearby farmer or contractor to do occasional heavy tractor work or trucking when necessary. We combine a little fun with work by using a small trail bike to make frequent, low-cost inspections of our plantings throughout the year. A trail bike or ATV 4-wheeler is also likely to help spark more interest in your enterprise among any young members of the family.

An impressive list of other items may prove necessary or useful, depending on where you live. Because all manner of animal pests take their toll of Christmas trees each year, when you plan your budget, consider that you may have to protect your plantings from deer, elk, the neighbor's cattle, or poachers by installing a tall fence, which can add a hefty amount to your ledger. You may also need irrigation equipment, fuel storage tanks, equipment sheds, and a tree-shaking machine for fast removal of dead pine needles. If you have a retail operation, items such as signs, snow removal equipment, and stands to hold your trees may add considerably to your debit column.

Don't overlook office expenses as you make out your budget. Advertising, insurance, taxes, postage, telephone, stationery, bank fees, and association dues may seem minor, but can add up to a large figure over a period of years, as can depreciation on such capital expenses as typewriters, files, computers, and cash registers.

As in most farm operations, careful attention to both production and harvesting costs, even the seemingly insignificant financial ones, can make the difference between success and failure.

CHAPTER FIVE

Which Tree
Should You Grow?

I often pass a gracious tree
  Whose name I can't identify,
But still I bow in courtesy,
  It waves a bough in kind reply.
I do not know your name O Tree,
  (Are you a hemlock or a pine?)
But why should that embarrass me?
  Quite probably you don't know mine.
      Christopher Morley (1890-1957)

We have learned that when our Christmas tree customers ask for a "pine" they don't always have in mind the long-needled varieties of the genus *Pinus* but often want some other needled evergreen, such as a spruce or fir. Even many natives of our rural wooded area do not know a fir from a hemlock and have no idea there are four different species of native spruce growing in our town. This seems strange because only a few generations earlier their ancestors knew the name of each tree and what each one was best used for — their livelihood depended on such knowledge.

Of course, you don't need to iden-
tify every evergreen species that grows across the country in order to grow good Christmas trees, but if you intend to create a successful plantation, and especially if your income depends on it, you should know the kinds best suited for growing in your region.

As I mentioned before, the selections of a site and the species of tree should be made together, because whichever choice you make first, the second must be compatible with it. Depending on your location, you may have a wide choice of trees that will grow well and be salable or you may be limited to only one species.

Evergreens grown for Christmas trees include the long needled pines: Austrian, red, Scotch, Virginia, white, and others; the short-needled firs: balsam, concolor, Fraser, and noble; the short-needled spruces: Colorado, Norway, red, and white; and another short-needled species: Douglas fir, which is actually neither fir nor spruce, according to scientific classification, but is commonly called a fir. Several species of evergreens without needles, such as arborvitae (white cedar), cypress, and juniper, are also used to a limited degree as Christmas trees in various parts of the country. Broadleaf evergreens such as holly and boxwood are widely used as holiday greens, but seldom sold as trees.

The three leading species currently grown in the United States and Canada for Christmas sales are Scotch pine, Douglas fir, and balsam fir. The latter two have long been cut from the wild and used as Christmas trees, but Scotch pine, a comparative newcomer from Europe, has only recently attained its great popularity.

If you are planning a small operation and selling only to the local trade, you may want to plant several varieties of trees so you can give your customers a choice. For a large wholesale plantation, however, you will want to grow only those species that you are certain are in demand and for which there is already an established market. If you are unfamiliar with the Christmas tree business, find out what other growers in your region are raising, and whether they have unsold trees left over at the end of the season.

Consumers often choose a Christmas tree that brings back memories of their youth, so tradition, in addition to availability and price, greatly determines what is in demand in various parts of the country. The firs and spruces are favorites in the Northeast, north-central states, the Appalachian Mountains, and southern Canada. Douglas fir is the first choice in

the West, but pines are more popular throughout most of the rest of the country, partly because they grow well in a wide variety of soils and climates. Your market need not necessarily be limited to what your own region prefers, however, because growers now ship millions of trees all about the country and even abroad.

Scientific Names, Species, Subspecies, and Varieties

Over the years, various common names for the same tree have evolved in different parts of the country. White spruce *Picea glauca,* for instance, is also called cat spruce, skunk spruce, and bull spruce. Botanists always use Latin-based scientific terminology because it provides unquestionable identification worldwide. *Picea glauca* is the same tree in Maine, Finland, Alaska, Siberia, or wherever it may be planted. *Picea* identifies the genus (spruce), and *glauca* the species (white). If several species are being described at one time, the genus may be subsequently abbreviated, as *P. glauca.* Subpecies may also exist. Black Hills spruce, a slow-growing white spruce from the northern plains, is officially *P. glauca Densata,* for instance, and one blue strain of Colorado spruce *P. pungens* is *P. pungens glauca* or Colorado blue spruce.

Many variations of these species and subspecies exist in nature, also. Horticulturists have discovered some exceptionally blue strains of Colorado blue spruce, for instance, and have named and propagated them. These named selections are called varieties. Koster's Colorado blue spruce (*Picea pungens* var. *'Koster'*) is one such variety that is propagated by grafting and sold for ornamental plantings. Scotch pine, grown throughout Europe for centuries, is another species that has spawned many different varieties including dwarfs and slim, upright growers.

Tree experts continue to search the woods of North America and Europe for strains that will make the best Christmas trees. Some of these are one-of-a-kind accidents of nature, but there are also wide variations within each species, depending on what part of the country and at what elevation the trees are growing. Evergreens acclimate to a region over the years and sometimes may not do well when moved to another area with a different climate.

Although most of the so-called superior trees have been nature's doing, a few result from human experimentation. Horticulturists cross various strains of the same species, and sometimes even two closely related species, hoping to develop new varieties that will combine the best qualities of both parents. Seedlings of crosses between Austrian and Japanese black pine are now being offered, for example, as are hybrids created from crosses of balsam and Fraser fir. Georgia-Pacific Corporation (Woodland, Maine, 04694) offers eleven different strains of balsam fir and three of Fraser fir.

Nurseries are now offering seedlings of these improved strains, and many Christmas tree growers are planting them instead of the common species. People who once considered the scientific names jargon, now use them religiously to be sure of getting the exact trees they want.

Even though the new introductions are superior to the species, growers are always on the lookout for the still undiscovered perfect tree that grows fast and is dense, with no shearing or spraying. They know that buyers want a tree with good color and good needle retention, but one which is still not priced beyond their budget. Many experienced growers believe that they have already reached top efficiency in their cultural practices, so planting better strains is probably the only way left to excel in the marketplace.

Don't get too excited, however, if you discover an outstanding tree in your woodlot. An exceptionally attractive wild tree often results not from gene combinations, but from soil nutrients or some other environmental condition. Trees growing in a rich field are likely to look far superior to those of the same species struggling in a worn-out pasture, for example. One especially fine specimen in our area created a lot of excitement, and growers clamored for seeds until a forester noticed that the tree was growing in the lush earth over the farmer's leach field! Trees grown in shade have thinner foliage than those in full sun, too. For many years woodsmen thought there were two kinds of balsam fir in our area — one with full, double needles and another with flat, single ones. Then someone noticed that the same trees often had full-needled growth on the sun-drenched top, and thin flat growth on the shaded, lower limbs.

Tree Species

Of the large number of evergreens that grow in the temperate zones of the world, only a few are widely used for Christmas trees. Many species are attractive but fail to have the qualities that homeowners want or growers find practical to raise. For the most part, pines, spruces, and firs supply most of the world's markets.

The Pines *(Pinus)*

Pines are popular with growers because they develop into salable trees rapidly, and grow well in a wide range of climates and in most soils, even though they prefer those that are somewhat acid. Pines are more drought resistant than most other trees, because after they are three or four years old they develop a deep-growing taproot. The trees need heavy annual shearing, which must be done during the short period in early summer when they are making their new growth. Because this shearing is time-consuming after the trees reach a few feet in height, be careful not to plant more pines than can be sheared properly in the required time.

The many shearings needed to force pines to grow into a compact shape also cause them to develop a thick stem that is heavy to handle. Their density also causes a large amount of dead needles to collect in the interior of the tree, and these have the unpleasant tendency of dumping onto the living room carpet as they dry out. The tree should be thumped on a solid surface by hand or shaken by machine to dislodge these needles before it is sold, but this is difficult if the tree is wet or snow covered.

Cones of the different species vary widely in size and shape, but all make good holiday decorations.

INSECTS: Sawflies, shoot and tip moths, spittlebugs, webworms, and weevils

DISEASES: Needlecasts, Scleroderris canker, rust, and twig canker

Austrian pine

Austrian pine *(Pinus nigra)*. Although this is an attractive ornamental, it is not the best choice for a Christmas tree planting because it is very susceptible to disease, and other suitable trees grow faster. Austrians make good "living Christmas trees" for planting out after the holidays, however, because unlike most pines they keep their lower branches even after reaching a large size. They can be grown over a wide area of the country, but will not withstand as low winter temperatures as the red, Scotch, or white species.

Jack pine *(P. banksiana)*. Has short (1-inch) needles and is able to grow in dry, sandy soil and exposed areas. The growth habit tends to be some-

The Pines *(Pinus)*

SCOTCH PINE

WHITE PINE

AUSTRIAN PINE

RED PINE

what loose, and the needles take on a yellow cast for the winter, so the tree enjoys only a limited popularity.

Japanese pine; Black pine *(P. thunbergiana or P. thumbergii)*. It withstands salt spray better than most evergreens and grows well in exposed seaside areas, but is better suited as a Japanese-type dwarfed tree in a landscape than for Christmas tree sales. Like Jack pine, it is planted only because other trees fail to do well in certain locations.

Monterey pine

Monterey pine *(P. radiata)*. This tree is native along the California coast where warm temperatures, abundant fog, and sandy soil provide the conditions it likes. Completely unsuitable for cold or dry climates, it grows rapidly into an attractive, dark green tree in a region where traditional Christmas tree species cannot be grown.

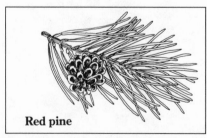
Red pine

Red pine; Norway pine *(P. resinosa)*. A North American native, so named because it once grew in large numbers around Norway, Maine. It has long, reddish needles, grows fast, and survives better than white pine in dry, unfertile soils. Not widely grown for Christmas sales, it nevertheless can be marketed well in certain areas.

Scotch pine

Scotch pine *(P. sylvestris)*. Scotch pine was originally imported from Europe for the fast reforestation of cut-over areas. It proved a disappointment as a timber and pulp producer, however, because it tended to be crooked and short lived, and had serious disease and insect problems. Resourceful growers started shearing a few and found they sold surprisingly well as Christmas trees. They can be grown well in milder climates where the cold-loving firs and spruces do not thrive.

Scotch pine seeds imported from different parts of Europe produce a wide variety of trees and the characteristics of some strains do not endear them to customers. The French, Spanish, and Greek strains have short (1-inch) needles and a good green color, but the trees grow slowly with thick, crooked stems. Spanish strains are especially susceptible to brown spot and Lophodermium diseases. Trees from central Europe, the Austrian Alps, Belgium, East Anglia, and

Great Britain have slightly longer needles, and grow fast with straight stems, but because of their fall yellowing, the trees need to be sprayed with a colorant before cutting to make them salable. German and Polish strains can be sheared into especially well-shaped trees, but develop even worse fall color. Even colorants fail to cover the intense yellow hue of trees from the more northern European countries. (Some enterprising growers, stuck with acres of canary-colored trees, tried to market them as "Golden pines" one year, but the idea failed to catch on with the public.) Although the Scotch pine seedlings and transplants now being sold by most nurseries are from the better strains, buyers should be on guard, because a few of the older, fall-yellowing kinds are still listed in catalogs.

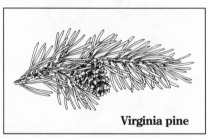

Virginia pine

Virginia pine; Scrub pine *(P. virginiana)*. This species was not considered valuable until recently, when growers discovered it could be shaped and marketed as a Christmas tree. Now it is widely grown in the southern states because it tolerates warmer temperatures even better than Scotch pine.

White pine *(P. strobus)*. The white pine has been valued as a timber tree for centuries, and when sheared

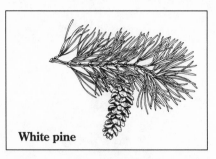

White pine

heavily it also makes an attractive Christmas tree. It is grown in the mid-Atlantic states mostly. Buyers admire its soft green color and woodsy fragrance, but sometimes object to the long needles that give it a coarse appearance. Growers like its fast growth habit and ability to do well in many different soils and climates, but don't always appreciate the heavy shearing that it requires. Some growers spray the trees with growth regulators such as Pro-Shear and report denser growth with less shearing. Pro-Shear is made by Abbot Laboratories (see Appendix).

White pine tolerates moist soils better than Scotch and red, but to grow well it needs a more fertile soil than either of those species. Its long, loose cones are in demand for decorating wreaths and garlands.

The Firs *(Abies)*

Firs, as well as spruces (see pages 35-38) are popular where they grow best in the cold sections of the U.S. and in southern Canada. They both have an attractive rich, deep-green color and conical shapes. They are more shallow rooted and less tolerant of drought than pines and like the coolness of a north-facing slope.

The Firs *(Abies)*

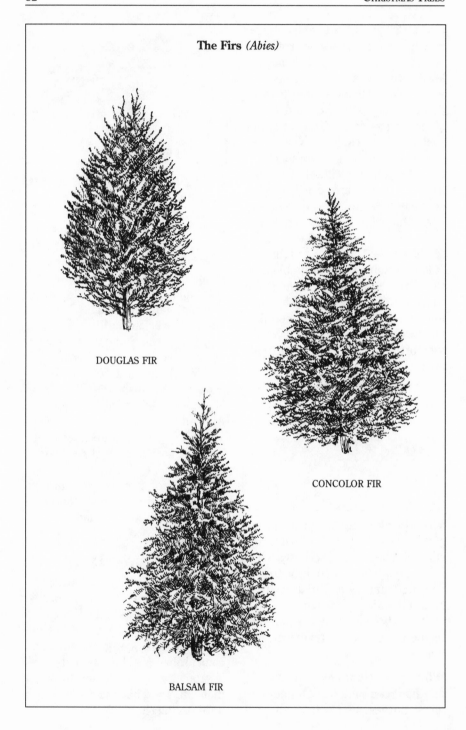

DOUGLAS FIR

CONCOLOR FIR

BALSAM FIR

Firs have needles that are flatter and softer in texture than those of the spruces. Their smooth bark becomes spotted with pitch blisters as the trees get older. The cones stand upright and burst apart on the tree rather than falling intact to the ground after opening, like pine and spruce cones. The trees seldom become wind-burned, but their early spring growth is sensitive to late frosts, so try to buy trees that are acclimated to your region and won't start growing too early. When the trees are young, they do not compete as well with weeds and grass as do the pines and spruces. Both the cut trees and greens enjoy good sales, as evidenced by the millions harvested each year in the northern United States and eastern Canada.

INSECTS: Balsam shoot boring sawfly, budworms, Cooley spruce gall aphids, gypsy moths, mites, Pales weevils, spittle bugs, woolly aphids

DISEASES: Cankers, Rhavocline needlecast, Uredinopsis rust

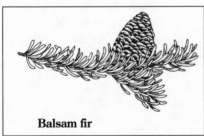

Balsam fir

Balsam fir *(Abies balsamea)*. This is definitely a cold-climate tree and does well only where winters are cold and summers cool. To many easterners a balsam fir is the only true Christmas tree. It is naturally cone shaped, with needles that are rich green on top and silvery white underneath. They last well, too, but their rich fragrance is the tree's outstanding quality. Balsam grows best on soils that are not too acidic, and it even does well on soils that are fairly moist, as long as they are supplied with nitrogen. Most strains need only three or four shearings before they reach salable size, and the shearing is faster and easier than on the coarser growing pines.

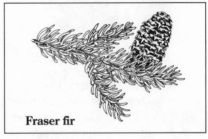

Fraser fir

Fraser fir *(A. fraseri)*. Most of what has been said of the balsam fir is also true of the Fraser; in fact, botanists don't agree whether it is a distinct species or a climatic variation of the balsam. Its needles are usually a bit shorter and the growth habit more dense than that of balsam. Fraser is more frost resistant, but also more susceptible to insect and disease problems, especially aphids, spider mites, budworm, and diplodia tip blight. Though it is less tolerant of wet soils than balsam, Fraser doesn't do well on extremely dry ones, either. It grows best on cool northern sites in zones 2 to 4, and grows naturally at altitudes over 5,000 feet in the Appalachians. The slender growth habit makes it appealing to buyers looking for a tree that is suitable for a small room.

Grand fir; Giant fir *(A. grandis)*. This tall-growing western fir is grown commercially as a Christmas tree in parts of Idaho and other states where winters are not too long or severe. It requires a cool, moist climate, but where conditions are right, it grows very fast.

Noble fir

Noble fir *(A. procera)*. This attractive tree grows naturally in only a small area in the Pacific Northwest and plantings in other parts of the country have not often been successful. For good growth, the noble fir needs a short growing season, high humidity but little summer rain. It does best at altitudes between 1,500 and 4,000 feet.

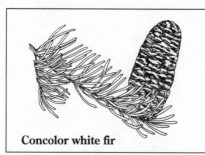

Concolor white fir

White fir *(A. concolor)*. One of the longest-needled firs, the concolor is sometimes mistaken for a species of pine. It makes a valuable landscape tree because its soft needles are bluish or greenish white in color. It is not widely grown as a Christmas tree because it is more difficult to shape than other firs and when sheared has the tendency to develop several new terminal sprouts each year.

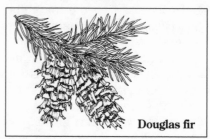

Douglas fir

Douglas fir *(Pseudotsuga Menziesii,* formerly *P. taxifolia)*. As the scientific name shows, the Douglas fir is not a true fir, but it does resemble the *Abies* genus in appearance with one exception: its large cones hang downward rather than stand upright as do those of the firs. It is grown as a landscape ornamental in much of the country, but in the U.S. Northwest, British Columbia, and the foothills of the Rocky Mountains it is the most widely grown Christmas tree. The tree grows cone shaped naturally, has a rich green to blue-green color, excellent needle retention, and can be sheared to a beautiful tight form. On the minus side, it has a low frost tolerance and does not compete well with grass and weeds.

Growers in the East often have trouble raising Douglas fir when they plant seedlings derived from trees grown near the coast, because these strains need a mild, humid climate with little summer rain. Seed from trees grown at higher elevations produce growing stock that is hardier and more tolerant of weather extremes. All strains require well-drained soil, good air drainage, and adequate soil nutrients.

Because Douglas fir strains vary considerably, growers should investigate sources and make every attempt to buy seedlings only from reliable nurseries.

The Spruces *(Picea)*

Spruce trees have stiff, square needles and rough bark, and they are prickly to handle. As a rule, spruces do not hold their needles well unless they are kept cold, so they should be cut as near Christmas day as possible and kept inside for only a short time. They are ideal as outdoor trees, however.

INSECTS: Budworms, Cooley spruce gall aphids, gypsy moths, mites, Pales weevils, spittle bugs, woolly aphids

DISEASES: Cankers, Rhavocline needlecast, Uredinopsis rust

Black Hills spruce *(P. glauca 'Densata')*. This subspecies of the white spruce is a native of the Dakotas. It grows naturally into a tight, compact tree with almost no shearing, so is good as a lawn tree. Its slow growth and poor needle retention make it impractical to raise it commercially as a Christmas tree, however.

Colorado spruce *(P. pungens)*. This tree is familiar to most people as a landscape plant, especially in its striking blue form. Although it is native to only a small area of the Rocky Mountains, it has adapted to a wide range of soils and to most of the northern part of the United States and lower Canada. In its favor are good form, good

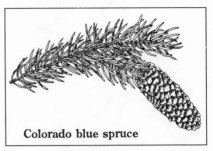

Colorado blue spruce

response to light shearing, and rapid growth after it reaches 3 feet in height.

Since it starts to grow later than most conifers, it is seldom damaged by late frosts. It is even more prickly and more difficult to wrap than the white spruce, however, and thorny to decorate. Although it is sold as a cut tree, and often billed as the aristocrat of Christmas trees, many growers have found that the most profitable way to sell Colorado blue spruce is as a potted or balled living tree.

Seedlings and transplants that have the best blue color are scarce and expensive, because only a small percentage of the seed, even from a good blue tree, will actually be blue; most will be various shades of green or blue-green. Not only are the blue seedlings scarce to begin with, but some seedling growers have been known to select out those with the best color and sell them to nurseries to grow for landscape plants, leaving those not as blue to sell to Christmas tree growers. Knowing this, some Christmas tree plantation owners raise their own planting stock from seed picked from a "blue" orchard — a planting of selected trees with the best color, grown especially to produce seed that will germinate the highest percentage of blues. (See

The Spruces *(Picea)*

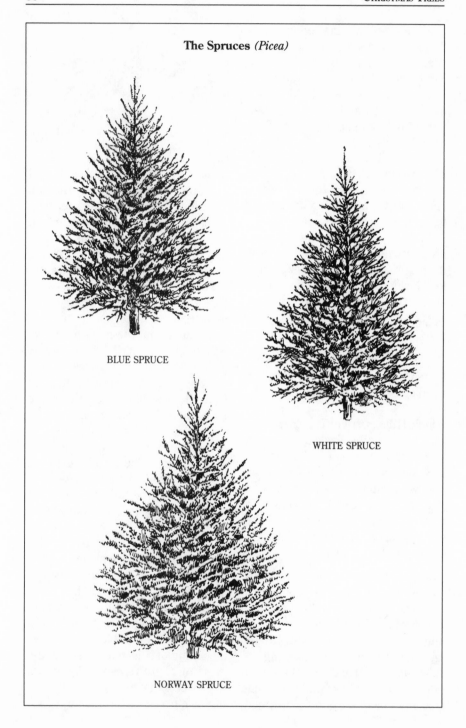

BLUE SPRUCE

WHITE SPRUCE

NORWAY SPRUCE

Appendix for seed sources.) When the seedlings are 4 or 5 inches tall, growers then move only those with the best color to a transplant bed (see pages 49), and discard the rest. After two years in the transplant bed, they grade them once more, selecting the bluest ones to plant into the field. This process is time consuming, and it means throwing away a huge percentage of trees, but it saves the labor and space of growing many poorly colored trees that won't sell.

As winter approaches, the powdery blue of the needles tends to fluff off, especially in windy spots, so a sheltered location helps to ensure good color at the time the trees will be cut. Additional nitrogen fertilizer applied at the beginning of the growing season usually improves their blue color.

A closely related species, Engelmann spruce *(P. Engelmannii),* has less spiny needles than the Colorado, and many strains of it also have a bluish color. It, too, is grown for both living and cut trees.

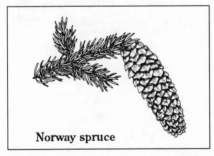

Norway spruce

Norway spruce *(P. abies).* This hardy, fast-growing tree does well in a variety of soils and climates and is widely grown in New Jersey as a Christmas tree. The tree needs heavy annual shearing if you want a tight tree, and in some areas its short needles and dark, somber color don't appeal to customers. It is prized for its huge, tight cones that are used for cone wreaths and other holiday decorations. In some regions, Norway spruce is used as a fast-growing, tight-to-the-ground windbreak that, as a bonus, produces loads of marketable cones.

Red spruce *(P. rubens).* A cool-climate tree, the red spruce is not widely used indoors for Christmas because of its reddish cast, objectionable odor when warm, and its poor needle retention. The boughs are often used for cemetery grave blankets and other outdoor decorations, however, and large trees are harvested to use in parks and malls and for other outside holiday displays. It is a hardy tree, but tends to windburn in exposed areas by the end of the winter. The browning usually disappears by late summer, however, so it is not usually a problem at the time it is harvested.

Serbian spruce *(P. omorika).* Many horticulturists feel that this tree has the most beautiful form of all spruces, so it is popular for home landscaping. Serbians have never been widely planted for Christmas trees because they are slow growing and their short branches tend to be droopy. Furthermore, they are not quite as hardy as the white, Norway, red, and Colorado spruces.

White spruce *(P. glauca).* At the country school of my youth, the boys

White spruce

green color and nice cone shape, and needs very little shearing.

Like red spruce it has poor needle retention and a strong odor in a warm room, but it is an excellent choice for outdoor decorations.

Other Species

Many other evergreens are used on a limited scale throughout the country, either because of local preferences, or because the common kinds don't do well in warmer regions. Afghan pine *(Pinus eldarica)* is grown in Arizona and New Mexico. Eastern white cedar or arborvitae *(Thuja occidentalis)*, red cedar *(Juniperus virginiana)*, and Cypress *(Cupressus)* all enjoy local popularity where they grow well, but all need to be cut early before they turn brown for winter.

If you are growing Christmas trees for the retail trade, don't be afraid to experiment with some of the kinds less commonly grown in your locality. There are always customers who are looking for the unusual, and supplying them can be good publicity for your business, as well as fun for you.

always brought in a large white spruce from the woods for our Christmas tree each year, because they always grew into the best shapes in the wild. The teachers always seemed pleased, too, until it began to give off its strong, woodsy odor in the warm classroom. Then we would get a stern warning about selecting a better-smelling, balsam fir next time. This threat was always long since forgotten before the following year, of course.

The tree grows naturally and at its best in Eastern Canada and the northeastern states. One of the hardiest evergreens, it needs a cold climate to survive. It thrives on a wide variety of soils, but grows more slowly when the soil is acid or wet. It competes well with weeds so it is good for growing in sod, has an attractive blue-

CHAPTER SIX

Sources of
Planting Stock

W hen we started growing trees, we were lucky to be able to deal directly with nature's nursery, pulling our seedlings from a nearby woodlot that was recently cut over, and planting them a few hours later in our fields and pastures. However, most Christmas tree growers aren't blessed with a source of high-quality wild stock and must buy seedlings or transplants from mail-order nurseries, although some of the larger plantation owners, in order to get the strains they want, raise their own seedlings.

Unlike ornamental shrubs, fruit trees, and perennials, which are commonly propagated by grafts, cuttings, divisions, and sometimes by tissue culture, Christmas trees are nearly always started from seed. Occasionally, however, outstanding trees are propagated by cuttings or grafts. Then a small number of these superior trees are grown together in a spot far away from other trees where they will be pollinated only by each other. These special "seed orchards" produce seeds that the propagators expect will grow into high-quality trees that will resemble their parents.

Buying Seedlings
and Transplants

Christmas tree nurseries (see Appendix) raise their planting stock by sowing seed in especially prepared beds. After two or three years, when the seedlings are well established, they are dug and either sold as *seedlings* or transplanted to new beds where they are grown for an additional two or more years before being sold as *transplants*.

As you read a nursery catalog or price list you will notice that planting stock is listed as (2-0), (3-0), (2-2), and so on. The first number refers to the number of years in the seedbed and the second to the years spent in a transplant bed. For instance, "Scotch pine (2-0), 8 to 12 inches" means they are two-year-old seedlings, from 8 to 12 inches tall; (3-0) are three-year-old

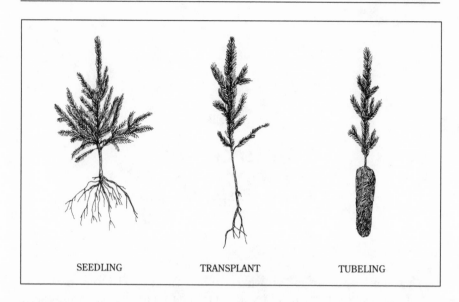

| SEEDLING | TRANSPLANT | TUBELING |

seedlings; and (2-2) are four years old, having spent two years in the seedling bed, and two more in the transplant bed. You may also find trees listed with three digits such as (2-2-2). These have lived two years in a seedling bed, been transplanted and grown two additional years, then transplanted once again to another bed and grown for two more years. Such a tree might be over 2 feet tall and require hand planting, but it could be harvestable within five or six years.

Seedlings have the advantage of being low in price, but their root systems are often small. This means it may take a year or more before they get well enough established to start growing after they are planted directly in the field. A two-year transplant (2-2) is the best choice for firs and spruces. The two additional years of growing time and extra growing space in the transplant bed allows them to develop a much larger root system; this better equips them to survive field planting and get off to a fast start. Pines can be an exception, however, and a good-sized, three-year-old seedling is often husky enough for field planting.

The extra time, work, and space involved in producing transplants naturally makes them more expensive, so some growers buy the lower-priced two-year seedlings, plant them about 4 inches apart in well-prepared beds of rich soil, and grow their own transplants. This method probably doesn't save money if you figure your labor as an expense, but having a supply of planting stock close at hand gives you more latitude in choosing when to plant. This is especially useful when it is necessary to replace trees that have died or been harvested.

Some nurseries also offer container-grown seedlings, sometimes called *tubelings*. These are grown in small plastic pots that are narrow in diameter, but several inches in length so there is ample space for root growth. They are usually started in a greenhouse and carefully nurtured so that after only one growing season they are often as large and heavily rooted as a small transplant grown outdoors. Container-grown plants become established quickly because the root ball is intact and undamaged by digging. They can be safely planted outdoors almost anytime after freezing temperatures are past in the spring and even in midsummer if they are kept watered. On the minus side, they are more expensive than bare-rooted trees, and removing those that are shipped in containers and disposing of the containers can be time consuming. Some growers report as well that tubelings grow more slowly than transplants.

Some states operate a tree nursery and sell seedlings and transplants for reforestation purposes. Because policies vary in different states, yours may or may not allow these to be sold to Christmas tree growers. Consult your extension forester about their availability if you are interested. The trees from state nurseries are usually priced lower than those grown by commercial nurseries, but since foresters think mostly in terms of timber production, the strains offered may not be the best choices for Christmas trees.

Wherever you buy trees, place your order early so you will be sure of getting the species, strains, and sizes you want. Most nurseries try to avoid having leftover planting stock, so they don't raise more seedlings than they expect to sell. When you order, specify the time you want them shipped, so you can be ready to plant as soon as they arrive. Make sure they are shipped by a dependable carrier, too. One of our grower friends was unlucky enough one year to have his planting stock stranded in mid-route for several weeks during a truck strike.

Most nurseries ship their seedlings promptly after digging and pack them well so there is no danger of their arriving in poor condition. Sometimes mistakes happen either in shipping or in transit, and if yours arrive dried out, or with the cartons broken open, file a complaint immediately both with the carrier and with the nursery.

Collecting Wild Seedlings

In areas where a good strain of evergreens grows wild, collecting your own planting material may be your best bet, especially if you want only a few trees. Look for wildings (uncultivated seedlings) growing near trees with good color and needle form. Recently cut-over areas, country roadsides, old logging roads, and neglected pastures are all good places to find spruce, fir, Douglas fir, and pine seedlings. Sometimes you can locate seedlings of species that are not native, such as Scotch pine, that have sprouted at the edge of planted forests or highway rest areas. Rural road crews often need to remove such trees growing in highway rights-of-way and the foreman will probably allow you to gather seedlings. In any case, before you dig or pull wild trees, always get permission from the owner if they are on private property, or from the proper authority if they are on public land.

Pulling wild trees is fun, and can be an enjoyable weekend family event or a job for the neighborhood kids. Although no skill is required and it is thus hard to do it wrong, it is important never to let the trees get dry. If the soil is moist, as it usually is in the spring, the trees come out easily with most of the roots intact.

We like to pull 8- to 12-inch tall seedlings in early spring as soon as the ground thaws and is still damp, and tie them in bunches of twenty-five or fifty, depending on their size. Then we quickly pop them into a moist burlap bag that we keep in the shade until we take them home. In a spot where heavy seeding has taken place, it is possible to pull several thousand in a day.

Most forest-grown wild evergreens have frail roots due to their crowded conditions, so you will get a higher percentage of survival if you first set the wildings in a transplant bed and let them grow there for a couple of years before planting them in the field.

Take exception to this rule when you dig larger trees rather than pull seedlings. If your planting is to be a small one, but you want the trees to mature as soon as possible, choose large, well-spaced trees of nearly any size. These can be carefully dug with a ball of earth and planted directly into your plantation without transplanting. Usually this can be done

successfully either in early spring or early fall.

One good argument for using wild trees is that they are already acclimated to the climate. Some of our neighbors who have bought seedlings from nurseries in states with growing seasons that are longer than ours, complain that the trees often start their new growth too early, and subsequently suffer damage from late spring frosts.

Raising Your Own Trees from Seed

Growing trees from seed is not only a practical way to start large numbers economically, but it is also fun. If you have the land, a source of good seed, and a little extra time, therefore, I hope you'll try it. If you want only a small number of trees, several hundred seedlings can be started easily in a small cold frame or grown in a few seed trays in a greenhouse.

The process is not difficult, but certain things have to be done at the right time or all your work will be wasted. Tree seedlings not only need more attention than crops such as peas and radishes, but special care is necessary to get the seeds to germinate well, stay free from disease, and grow quickly to transplanting size.

If you decide to start your own trees, you can either buy seed from commercial seedsmen (see Appendix) or collect your own. If you buy, try to select a firm that offers seed from good strains of trees grown in a climate similar to yours. Most companies will furnish a list of their seed sources.

Collecting Seed

Seed gathering is practical only if you have access to a good strain of trees. Trees seldom produce cones more than once every two to four years and less frequently if frost damages the spring blossoms. Be on the lookout for heavy cone seasons and gather extra seed to save for future years (see page 44).

Conifers produce their seed between the scales on cones. The seed itself is rather small, but nature has outfitted each one with a large wing, so that when the cones dry and open on the tree, the wind can carry the seed great distances. Cone sizes vary from the tiny (3/4-inch) hemlock ones to those of the sugar pines, which can be 20 inches or more in length. I have often thought that walking through a grove of sugar pines during cone dropping season would be quite exciting.

Seed sources may be anywhere: roadsides, parks, wild or planted forests, or even the home landscape. The best time to collect cones for seed is in late summer and early fall when they are fat, covered with pitch, and about ready to open. To be sure of the right time, watch a group of experts — the squirrels — and start gathering as soon as they do. Since you can't run up a tall tree as easily as they, you may have to pick yours from shorter trees or, even better, collect them from a lumber or pulp-cutting enterprise that is taking place during ripening season. There you can easily gather bushels of cones in a short time from the leftover tops. Unless you remember your gloves, you'll get your hands very sticky in

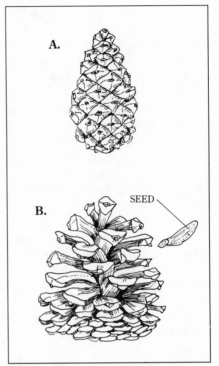

Conifers produce their seeds between the scales on cones (A). *As the cone matures and the scales open, the winged seed is released* (B).

the process.

Spread out the cones one layer deep on elevated screens so the air can circulate around them. Place them in the sun or in a warm, airy attic or open garage where you can protect them from mice and squirrels, and let them dry until the cones open. If you dry them outdoors, bring them inside at night and on damp days so they won't collect moisture.

After the cones have opened shake the seeds out. Dry the seeds a day or so longer, then rub them gently over a screen. The screen mesh should be of a size that will allow the seeds to fall through the mesh as the wings are scraped off. Although you can plant seed with the wings still attached, as nature does, the percentage of those germinating is likely to be smaller because the wings gather moisture, encouraging rot and other diseases. After removing the wings, as a further precaution against disease, clean the seeds by winnowing — pouring them from pail to pail in a light breeze.

It is best to plant tree seed in the fall, as nature does, but if that is not possible or if seed is to be saved for future years, seal it in a glass jar or plastic bag and store it in a cool, dry place. The seed of most species remains viable for two to four years, but older seed will not always sprout as well.

For spring planting, the stored seed should always be stratified for a few months just previous to planting in order to germinate well (see box on page 47). Planting in the fall stratifies the seed naturally.

Planting Seed

When you need a certain number of trees and plan to raise them from seed, it is always a good idea to plant far more seed than you think you may need. Only about twenty percent of balsam fir seed germinates at all, as a rule, although most have a much higher rate. The amount of seed for 100 square feet varies from 20 ounces for balsam fir to 6 for Norway spruce. Most of the pines need about 6 ounces, and Douglas fir about 12.

When you need only a few hundred trees or less, the best way to start them is in seed flats. Temperature and watering are easier to control, so there is less danger from damping-off diseases that strike young seedlings just after they have sprouted and are growing well. Since these diseases always exist in farm and garden soils, fill the flats instead with an artificial soil mix, such as Metro-Mix or Pro-Mix, which is available at most garden centers and farm stores. Be sure not to buy topsoil or potting soil, which are usually not sterile, and keep the artificial mix uncontaminated by not infecting it with tools or fingers you have had in garden soil. If you plant seeds in the fall, put the flats in a cold frame, in a root cellar, on an unheated greenhouse floor, or in a cold basement, or cover them with several inches of evergreen boughs and leave them outside in a sheltered place. Keep them moist, and protect them from mice and squirrels by putting a few mothballs into the flats.

When spring comes, move them to a warm, lighted area, but keep them out of direct sunlight and protect them from freezing. Sprinkle slow-acting fertilizer such as dried manure or Mag-Amp over the top of the flat. Tree seeds need moist, semi-warm, shaded conditions to germinate, but unfortunately, damping-off thrives under those same conditions. One's natural inclination would be to give them lots of warmth and sunlight, conditions that would help protect flower and vegetable seedlings from disease; but sprouting tree seedlings don't like either direct sun or too much heat. To help control disease, keep the soil mix moist, but try not to let the tops of the seedlings stay wet for long periods. Water early in the morning so the tops can dry out quickly. If disease strikes, spray immediately with a fungicide and continue spraying weekly until midsummer.

Outdoor Seedbeds

For large numbers of seedlings, plant the seed in outdoor beds in fall or early spring. A plot 4 feet wide is convenient to reach across, and at a rate of fifty seedlings per square foot, a bed 20 feet long can accommodate approximately 4,000 well-spaced seedlings. A light, sandy growing medium is best, so if your soil is heavy, mix in generous amounts of sand and peat moss. Prepare the seedbed by tilling it thoroughly so it is well pulverized and as free of weed roots as possible, then spread dried manure (at the rate of 1 pound per 8 square feet) over the bed and rake it in.

To be sure that the bed will be free from weed seeds and disease, sterilize it with Vapam or a similar disinfecting product (available at farm or garden supply stores) before planting. Follow the directions on the package carefully.

You may plant by scattering seeds sparingly over the bed, but it is easier to space them evenly if you mark out rows 2 inches apart and place the seeds in them. Cover them with 1/4 inch of clean, sifted sand or perlite and scatter a few mothballs throughout the beds to discourage rodents.

If the seeds are planted in the fall, mulch the beds with several layers of evergreen boughs or several inches of leaves. Remove this mulch in the spring after the ground is no longer freezing at night. Since conifer seeds sprout well only in fairly deep shade,

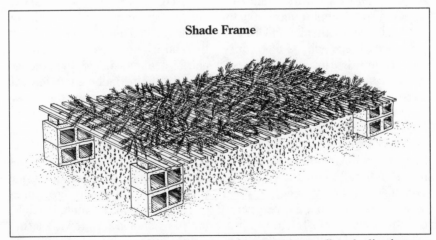

Shade Frame

A frame of snow fencing covered with evergreen boughs makes excellent shading for young seedlings.

you'll need a frame to support some type of shading material. (See illustration.) Snow fence or lath held up by posts or cement blocks about 18 inches above the beds works well. Cover the fence or lath with evergreen boughs — the ones you used for winter mulch will be fine — commercial plastic shading, two layers of burlap, or some other material that will provide a dense shade, but still allow rain and waterings to penetrate.

The seedlings under the shade need gentle treatment since they are so tiny and fragile. Unless it rains, water them about twice a week until they are an inch or two tall and growing well. Then water them only when the ground appears dry. Use a sprinkler can or hose with a fine spray in order not to damage them. To encourage fast growth, add a weak solution of liquid fertilizer (half of what is recommended on the package) to the water once every two weeks. Discontinue liquid feeding after the first of August so all growth will stop and the plants will have ample opportunity to harden up before the first fall frost.

Check the seedlings often during this time, because this is the period that damping off is most likely to strike. We once had a large bed of beautiful seedlings collapse and die in less than a week because we were too busy to keep watch. Spray a fungicide at the first drooping of the plants. After all this work, you don't want to lose your investment.

If you have used evergreen boughs for shade, most of the needles will have dried and gradually fallen off by August. This will allow increas-ingly greater light to reach the seedlings, which they now need, and the needles will provide a nice mulch. If you've used other forms of shading, remove most of it, except for the lath or snow fence, which should be left in place to furnish light shade for the rest of the season. As the days shorten in late fall, remove these, too, so that as soon as the snow starts to fall, it can cover the little seedlings. After the ground begins to freeze, protect the plants from frost heaving by covering the seedbed with evergreen boughs or an insulating foam blanket, available from nursery supply houses and garden stores.

Remove the boughs or covering as soon as the ground stops freezing the following spring. As the seedlings start to grow, thin them so they will be about 2 inches apart in each direction. Give them liquid fertilizer weekly during early summer, and water them whenever the weather is dry. Pine seedlings need no shade the second year, but replace the lath shade over spruce or fir seedlings, and leave it on until late summer. At the end of the season most spruce and fir seedlings should have reached a height of 3 to 8 inches, and they will be ready for transplanting the following spring. Any smaller ones can be left in the bed to grow there another year. Pine seedlings are more vigorous and will probably be 6 to 10 inches tall. These will also be ready for spring transplanting. The strongest ones can be planted directly into the field, but most should be grown another year or two in a transplant bed.

Seed Stratification

When the seeds you have bought or collected cannot be planted in the fall, seal them in plastic bags or glass jars and store them in a cool, dry place. Never plant stored seeds directly in flats or the ground, however, because they need a chilling period in a dark, cool, moist place for several months before they will germinate well. This process, called seed stratification, is provided effectively by Mother Nature, who does her planting in the fall.

To stratify seeds that will be sown the following spring, place them between layers of moist sand or sphagnum moss in a mouse- and squirrel-proof container. Dig a hole in the ground and bury the container beneath several inches of leaves where it will remain over the winter. It will be easier to separate the seeds from the moss or sand if they are placed between sheets of fine screening that are stapled together to make an "envelope." You can plant any of your stored seed in the fall without stratification, because fall planting takes care of that neatly, but if you plant in the spring always stratify it first or germination will be poor. Treat only the amount you expect to plant and keep the remainder in a sealed, dry condition.

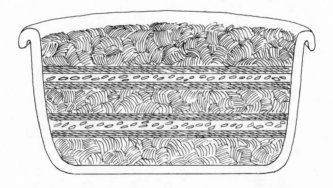

 SPHAGNUM MOSS   SCREEN   SEEDS

To stratify seeds, place them between sheets of fine screening, which are in turn placed between layers of moist sand or sphagnum moss in a mouse- and squirrel-proof container.

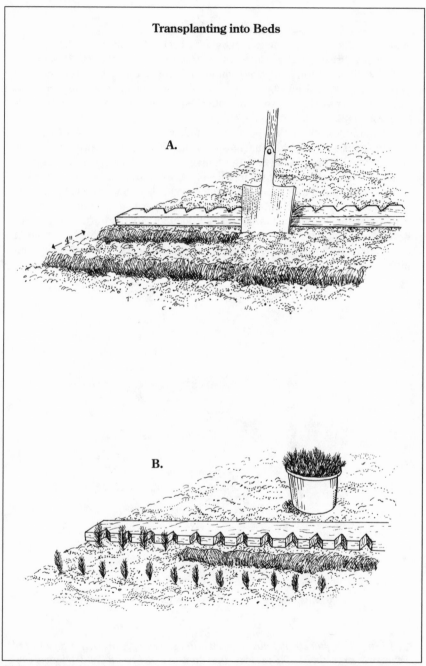

Transplanting into Beds

A.

4"

B.

A. *Use the straight side of a planting board as a guide to cut a trench across the bed.* B. *Set the seedlings in the trench, placing one seedling at each space on the notched board.*

The Transplant Bed

Although a few growers still plant seedlings directly in the field, to save losses and to avoid tying up the large fields for several extra years, most prefer to transplant their seedlings and let them grow an additional two years in a transplant bed. Here they can develop a much sturdier root system and be better able to stand the shock of growing in the less favorable environment of the field. The extra work of caring for the trees in a transplant bed for a couple of years will be well worth it whether you buy the seedlings, pull them from the wild, or grow your own.

Choose a sunny spot for your bed, and till and fertilize the soil as thoroughly as for a seedbed. It can be any length and width, but one measuring 5 or 6 feet in width seems to work well. If the seedlings are spaced 4 inches apart, a bed 6 feet by 18 feet will hold nearly 1,000 trees. Transplanting can be done in late summer if the trees are kept watered and shaded, but spring is a better time because they become established faster then and will need no shade.

Choose a cloudy day, if possible, and never let the seedlings dry out. Dig a trench across the bed with a shovel and set in a row of seedlings. Space them 3 or 4 inches apart in the rows, which are also spaced 4 inches apart. At a 3-inch spacing twenty-five seedlings will fill a row across a 6-foot bed. Then firm the soil back against the roots, making sure the trees stand upright. If you are working alone, dig and set only a few rows at a time, so the roots won't dry out, and water each batch thoroughly after planting. If you plant a lot of seedlings, a notched planting board cut to the width of the bed will greatly speed up transplanting (see illustration). With one person digging the seedlings and another planting with the notched stick, 4,000 seedlings can easily be transplanted in a day.

After planting, mulch between the trees with wood shavings, evergreen needles, buckwheat hulls or a similar organic material, and follow the watering and fertilizing recommended for second-year seedlings. Keep the area free from weeds.

The first year you will notice only a little top growth on your transplants, but root growth will be substantial. The second year, both roots and tops should grow heavily and the following spring they will be ready to plant directly into the field.

CHAPTER SEVEN

Choosing a Cultural Method

I t will save a lot of future readjustment, money, and work if you choose the cultural practices you plan to follow for the next ten years before you plant your first tree. For example, decide early whether you wish to use chemical fertilizers, pesticides, and herbicides, or to grow the trees as organically as possible.

Ten different Christmas tree farms will most likely exhibit a wide variety of planting, shearing, and weed-control methods, and each one might appear to be producing good trees. If you visit each and observe them closely, however, you may see that while each has its merits, each also has some negative facets that you should understand before making a decision about how to grow healthy, handsome trees.

The Sod Method

Keeping the land entirely in sod was the method used almost exclusively in the early days of Christmas tree culture. To me and many others,

because the trees and grass coexist much as they do in nature, this method seemed the logical way to grow trees, and it still is the favorite of a great many growers. To encourage good tree growth, extra fertilizer is applied from time to time, so both grass and trees are well fed.

Planting a small tree, even a transplant, into sod places the tree at a disadvantage, however, since it has to compete with well-established vegetation for nutrients and moisture. To give the fledgling a better chance, most growers try to kill or suppress the grass and weeds around the new tree either before or after planting. Growers used to scalp the ground with a grubbing tool called a mattock before inserting the seedling. Now they often either till the ground, use an herbicide, or, in the case of small plantings, apply a mulch. As soon as the tree is three or four years old, it can usually compete quite well with a grass cover.

The advantages of sod culture

are many, and it appeals especially to growers who prefer organic and natural farming, since it makes it possible to raise trees without chemicals. Bare soil erodes and loses organic matter quickly when exposed to the sun and wind; it loses further nutrients and topsoil when washed by heavy rains or snow melt. Soil that is covered by grass is not only better protected, but when it dies down each winter, it adds humus and nutrients to the soil, gradually improving it. This buffering action also helps avoid some of the problems caused by excessive use of chemical fertilizers and pesticides. The deep, rich topsoil the early pioneers found when they first came to this country was formed by many milleniums of such soil building.

Grass sod keeps tree roots cool in summer and protects them from extreme cold in winter. Because sod absorbs moisture, rain soaks into the soil rather than runs off as it would on bare earth. A covering of grass also helps prevent the soil compaction caused by walking or driving equipment over it. Sod also encourages earthworm activity, which helps keep the soil loose, thus allowing tree roots to grow more easily.

Although some tree producers let grass and weeds grow unchecked and cut out only excessive brush growth, most agree that rank herbaceous growth is an impediment to good tree development. When weeds are allowed to grow, they not only compete with the trees for moisture and nutrients, but are also likely to deform the lower branches of the tree by crowding and shading them. High grass and weeds also hide rocks and stumps that can cause damage to mowers and other machinery. Large colonies of tree-nibbling mice and rabbits often live happily amongst any thick thatch, and woodchucks, gophers, and moles dig holes in it. For these reasons, few growers manage their plantings on a totally live-and-let-live policy, but instead keep excessive growth under control, especially during the early life of the trees. A few fastidious people with small plantations mow every week and clip carefully around each tree with a string mower or by hand, keeping their grass like a lawn. Since this is expensive and time consuming, most growers mow only once a month or even just two or three times during the summer. This keeps weed growth from getting tall enough to harm the trees. The grass clippings make a good mulch and add humus to the soil as they decompose.

The Bare Earth Method

At the opposite end of the spectrum is a growing program by which the land is maintained completely bare of everything except the trees. This is accomplished either by herbicides or sometimes by cultivation, if the plantation is small. Those who prefer this method feel that both the nutrients they apply and those that already exist in the soil should all be available to the tree, and any ground cover offers too much competition. Such plantings are impressively neat, but I'm one of those who cringes at letting so much topsoil lie naked to the elements. In my experience, erosion causes a rapid loss not only of nutrients, but also of valuable organic matter. Without humus, as I have said, light soils dry out quickly and those with clay turn to deep mud in wet weather and bake hard in dry seasons, making good root growth impossible. Humus also serves as a soil buffer by absorbing excessive amounts of moisture, lime, and nutrients, later releasing them in small amounts as they are needed by the plants.

You should not consider the bare-earth method if your planting is on a slope and the soil is of the type that will wash badly during hard rains. Erosion not only destroys valuable topsoil, but can also result in deep ditches that are difficult to repair.

Combination Methods

Many growers use a combination of these two practices: the advantages of a sod cover with minimal competing growth. Grass is allowed to grow over most of the area, but before planting, a strip where the trees will be set is tilled to kill the grass and weeds and to loosen the soil. Instead of tilling, many growers use an herbicide to clear either a weed-free circle around each tree or a narrow strip along each row of trees. Most growers prefer the latter because the chemicals are easier to apply along a row. This method requires more herbicide, however, and opens a long stretch of exposed soil that invites erosion. Herbicides should be used the fall before spring planting so they won't damage the new trees. With either method, the sod strip between the trees is usually mowed at least once during the season.

Raising Living Christmas Trees

"Living" Christmas trees — those grown and sold with their rootball wrapped in burlap or plastic — appeal to practical folks who hate to buy a tree and then throw it out after only a few days of use. They like to enjoy a tree indoors during the holidays and then plant it into their landscape. Living trees also are appreciated by ecologically minded people who feel that cutting and throwing away a tree each year is a waste of natural resources.

The demand for live trees is limited, so be sure of your market before you plant the North Forty. Many people like to keep a tree in their homes for a week or more, but a dormant tree will start to grow if it is kept in a warm room for more than three or four days. Where the ground is frozen, it is impossible to plant, unless the hole has been dug earlier and there is plenty of mulch to cover it with after planting. In areas where there is deep snow, a balled tree can be buried in snow for the winter, but it should not be allowed to dry out. There is also a limit to the number of large evergreens most families need before their exterior decorating is complete. A further discouraging factor is that much of the demand for living trees is filled each year by landscapers, nurseries, and garden centers who market their surplus Douglas fir, blue spruce, and other ornamental evergreens at Christmastime.

Growing living trees can still be economically feasible, however, because the trees sell at high prices. Balled or potted trees are expensive to produce though, since the trees to be dug must be root pruned (have their roots cut around with a shovel) every three or four years to keep the soil ball compact. This, plus the time-consuming digging and wrapping, must all be included in the price. Transporting the heavy soil ball also adds to the cost, and because of their weight, live trees are seldom more than 3 or 4 feet tall. Also, consider that any unsold trees must be replanted.

In addition to the balled-and-burlapped Christmas trees, living trees are also sold in large pots or tubs. Sometimes plantation trees are dug and replanted in containers just previous to being sold, but often they have been grown in pots for their entire lives. Plants in pots usually grow much faster than those grown in the field, because watering and feeding can be more carefully controlled. Another advantage to pot culture is that a great many trees can be grown in a small space. Seedlings can be set in 6-inch pots initially, then transplanted to larger containers and spaced further apart as they grow. Besides making efficient use of space, growing in pots eliminates the labor of root pruning, digging, and wrapping. Also, since soil is usually purchased for potting, no valuable topsoil is lost from the field as it is when trees are dug.

There are disadvantages, though. Large pots are expensive, and pots must be watered daily during three seasons, a time-consuming process. Automatic irrigation systems are available, but the initial investment is high, and they must be monitored daily to make sure none of the small watering tubes to each pot have become plugged. Watering daily also leaches the nutrients from the pots rapidly and these must be replaced frequently. A slow-acting fertilizer, combined with regular doses of a liquid fertilizer during the first part of the summer, is commonly used to sustain good growth.

If you decide to grow living trees, be sure to raise only varieties that can be used as good yard trees in your area. Some of those most commonly grown are blue spruce, Douglas and concolor fir, and Austrian, white, and red pine. All will eventually become too large if grown close to the house, so tell your customers that if they plant them as foundation ornamentals the trees will need to be sheared heavily every year forever.

Although choosing a cultural method is an important decision, and the wrong choice may mean money and time wasted, it is possible to change from one practice to another during the life of the trees. You may want to experiment with small areas to see what works best with the species of trees you are raising, your soil type, and your climatic conditions, and modify your cultural practices accordingly.

CHAPTER EIGHT

Preparing the Soil, Planting the Trees, and Aftercare

I f you have a lush, fertile field to start with, you may not need to tackle any soil improvement before planting — and you can count yourself very lucky. The condition of most soils is not that good, though, and usually some sort of initial treatment is necessary for best growth. As with most growing practices, there are many different ways to improve your soil and you will have to decide what is the best for you.

Some growers prepare the area as if they were planning to plant potatoes — by tilling and fertilizing the plot, adding lime, if necessary, then planting oats, buckwheat, or millet. In the late summer or early fall, they till the grain crop into the soil before it ripens and plant their trees the following spring. This method adds both nutrients and humus to the soil.

Others prefer to grow their trees in sod. Before planting, they test the soil, then apply whatever nutrients and lime the test results recommend. Doctoring up the soil is easier to do before the trees are planted, because you can cover the area more evenly whether you spread by hand or use a spreader. If your planting is large, you may want to ask a fertilizer company to analyze your soil and apply the proper treatment with a truck spreader. Lime and fertilizer are heavy, so power spreading saves not only time but also a lot of back strain, while at the same time guarantees the most uniform covering.

Planting

Spring is the best time to plant tree seedlings because they then have an entire growing season to become established before facing the ravages of winter. It is better to set out transplants in spring, too, because the ground is more likely to be moist then, but late summer planting is practical also if the weather is not too dry. The larger root system of transplants enables them to become well established before the ground freezes and they can then get off to a quick

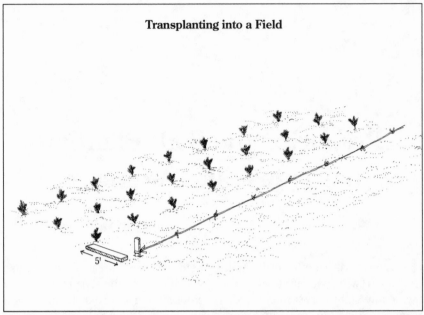

Transplanting into a Field

For speed and accuracy, use a planting string with knots at intervals of 5 feet and a 5-foot board when transplanting seedlings into their permanent growing locations.

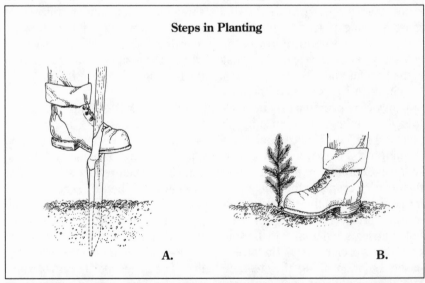

Steps in Planting

A.   **B.**

To transplant a young seedling, (A) drive the spade straight down into the earth and move it back and forth to open a wide slit, insert the seedling 1/2 inch deeper than it originally grew, and then (B) press the soil firmly back around it.

start in the spring.

Speed is important in planting. I used to tell my helpers to think of the trees as trout and never leave the roots exposed to air for more than a short time or they would die. We always kept the pile of unplanted trees thoroughly soaked and covered, and carried the ones we were planting in a pail of water. Since the soil was still moist from melted snow and spring rains, we didn't need to water them, so the results were usually good.

Although nursery-grown seedlings and transplants are better rooted than those pulled from the wild, they are just as perishable; but for some reason, it is difficult to convince some planters of this fact. I have been astonished to see growers unpack an entire shipment of several thousand small trees and leave them lying in the sun and wind until the planter was ready for them. Those same growers are the ones who usually complain about poor planting stock.

Mail-order trees are usually tied in bunches of fifty, packed with their roots in moist sphagnum moss and wrapped in burlap that has an outer covering of plastic or coated paper. The plants are likely to be dry after their trip, so as soon as they arrive, remove the outer wrapping and spray or pour water over them. If you cannot plant them immediately, store them in a basement or other cool place protected from sun and wind. Keep them moist and they will stay in good condition for several days. If you can't plant them within a week, each bundle should be untied and the trees spread out and temporarily

planted or "heeled" into trenches dug in a garden or other recently tilled area. Keep them watered and lift them out as needed for planting. Be sure, however, to get them into the field or transplant bed before any new sprouts appear.

If your climate tends to be dry at planting time, you can help keep the trees supplied with adequate moisture if, before planting, you dip their roots in a slurry made of a water absorbent, such as Supersorb (sold by Aquatrols Corporation of America; see Appendix). According to the company, the chemical is able to absorb and later release up to 200 times its weight in water, and it continues to work for up to two years.

Spacing of Trees

Growers don't want to waste land by planting too sparsely, but far too many forget that overcrowded trees can't grow well. The 4-by-4 foot spacing that was once recommended does not provide enough room, even though trees don't all grow at the same speed. Trees that will be cut when they are 5 to 7 feet tall are usually planted 5 feet apart in both directions. Those that will be grown to larger sizes before harvesting need more room, and smaller, tabletop trees obviously require less. If you plan to use a power mower to control weed growth, the space between the rows must be wide enough to accommodate it, even after the trees have increased in size. Of course, if you are aiming to grow a 30-foot specimen for a city park, you will have to allow far more spacing—and about thirty years

---

Planting Strings

Planting strings or wires are a big help when setting trees by hand; tree equipment companies supply them in different lengths with marks spaced along the string at various intervals. The string is stretched along a row and a tree planted at each mark. If the first trees in each row are planted in a straight line, all the trees will be in straight rows up and down, back and forth, and diagonally (see illustration). This type of planting not only looks nice, but also makes it possible to mow between the trees in several different directions.

I like being a do-it-yourselfer, so I make my own planting strings out of baler twine. I cut the twine into 5-foot-2-inch lengths and tie them together. (The extra two inches allows for tying the knots.) A 200-foot length is convenient to use if the field is fairly level, and when moved forty times it covers about an acre. By having a 5-foot measuring stick at the end of each row, the string can be moved from row to row quickly and accurately.

---

of growing time.

Unless your operation is very small and you do everything by hand, leave plenty of space for roads throughout your plantation. You won't want to drag the cut trees long distances at harvest time or carry fertilizer very far to the growing trees. Roads also provide fire protection. In a large plantation, ideally there should be a road every fifteen to twenty rows of trees. Omitting two or three rows makes a road of adequate width, unless your equipment is very large.

Hand Planting

Hand planting is necessary in many situations: on rough stony pastures or steep hillsides, for replacing trees that have died, and for interplanting between trees that will soon be harvested. Even on smooth, level fields, if you intend to set out only a small number of trees each year it is usually more practical to plant by hand than to buy or rent a machine.

When planting by hand, use a spade or planting tool (see illustration). The planting implement should be driven straight down into the earth and moved back and forth to open a wide slit. It is tempting to dig into the earth at a slant, particularly if you are using a spade, but if a tree is inserted at this angle the roots are left close to the soil's surface where they can dry out quickly. The hole should be large enough so all the roots will fit in completely. Any that are exposed to the elements will quickly die, and this will weaken the tree greatly. The tree should be set about a half inch deeper than it grew in the transplant bed.

If the soil hasn't been tilled before planting, it is important that it be loosened a bit with the shovel or planting tool before the tree is inserted. Otherwise the roots will have difficulty in sprouting in the compacted soil, and in dry years, the tree may die because it is unable to obtain moisture. Even if weather conditions are good, poor root growth will also mean poor top growth and the first growing

Planting Machine

season will be completely lost.

A gasoline-powered auger does an effective job of loosening the soil and makes it easy to insert the tree speedily. Unlike most other Christmas tree tools, the auger can be used at other times of the year for jobs such as digging holes for fence poles or for ice fishing.

Because of my lazy streak, wherever the ground is not level I like to start at the bottom of the hill and plant going up, rather than going down a slope. That way I don't have to bend over as far to set in the trees!

Sometimes the roots of small trees are long and stringy, which makes it difficult to plant them rapidly. If you cut an inch or two of roots off the bottom before you untie each bundle you can speed things up considerably. Don't worry about hurting the tree: Root pruning also stimulates

it to start growing faster.

By using a planting string, one person alone can plant 600 to 1,000 trees a day; two working together can easily do three times as many. The more experienced planter usually does the planting and the helper follows to make sure each tree is well inserted and to step on the loose soil around each one so no air is left near the roots to dry them out. Three people can also make a good team. One opens the slit with a spade, auger, or planting tool, one puts in the tree, and the third presses the soil back around it. Trees can be planted nearly as fast as the team can walk.

Planting Machines

If your operation requires setting out thousands of trees annually, owning a tree planter may be practical. There are many models of tree-planting

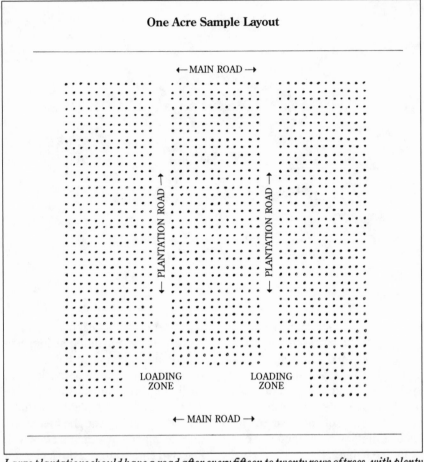

One Acre Sample Layout

Large plantations should have a road after every fifteen to twenty rows of trees, with plenty of room to load trees and turn heavy equipment.

machines that attach to tractors, all greatly improved from the tobacco planters the early growers had to use. With some, and a three-person crew, it is possible to set 10,000 trees a day. If you choose one of the larger machines, make sure that you also have a tractor powerful enough to operate it. Unless your operation is large, it may be more practical for you to contract for someone to do your planting. A planting machine is fast and sets the trees evenly spaced in straight rows, but it must be properly adjusted and operated by skilled workers. It should open a slit in the soil large enough so all the roots of the tree can be speedily inserted, and then the soil must be pressed firmly around the roots to prevent any air pockets that would dry them out. Since this procedure is so critical to the survival of the trees, many growers hire someone to walk behind the planter to ensure that each tree is upright and well anchored in, and to step on the soil around each one. This pressing also leaves a slight depression that will collect future rainwater.

If you hire a contractor, make sure the small trees are on hand when he arrives, because tree planting crews like to go at full steam once they start and any delay is costly. If you do not oversee the planting process, be certain the workers are familiar with your plans, because it is difficult and expensive to correct mistakes later. Indicate where you want to leave roads and other unplanted areas for parking vehicles, tying trees, and loading trucks during future harvests. Once I saw a 40-acre field, newly planted with Colorado spruce, without a single road or cleared space in the entire lot! I wish I had gone back later to see how the trees were taken out. With considerable difficulty, I'm sure.

Aftercare: The First Years

The first year after planting is the most critical in the life of a new tree. If conditions are favorable, a transplant can make considerable growth in the first year, but if they are not, it will have difficulty fending for itself. During a dry season the small plant may be unable to get enough water and nutrients, since the roots can reach only a few inches into the soil. Even if a tree survives this kind of stressed condition, it will be especially vulnerable to insects and disease later.

Since the growing season of evergreens is early and lasts only three or four weeks, whenever a tree adjusts too slowly to its new surroundings, no growth at all may take place the first year and a whole growing season is lost. If you can plant in the midst of a rainy period, it will be worth the trouble of getting wet. Not only are the little trees less likely to dry out, but they can make quick use of the nutrients in the moist soil and start growing at once. Plan to water or irrigate your trees for the first few weeks after planting if Nature refuses to do her part.

Keep a lookout for any harmful insects or diseases that have discovered your trees (see pages 85-96). The fledglings need everything in their favor.

The spring after planting, check the lot for winter damage, replace any dead trees, and straighten up those that were bent over by winds or heavy snow. Cut off any extra tops that may have sprouted, and reset any trees that have been partially heaved out of the ground by frost.

Mowing and Mowers

If you haven't killed the grass by tilling or by herbicides before planting, begin mowing at once to keep grass, weeds, and brush from crowding the new trees and stealing their moisture and nutrients. Most large-scale growers like to use heavy-duty, rotary-type mowers because they are easy to control and the blade can be readily sharpened. Others prefer those with a sickle cutting bar in front. A walk-behind, wheel-mounted, heavy-duty string mower can be practical for the small grower because not only does it cut weeds, grass, and fresh brush growth, it needs no sharpening; and if it hits a stone, there is no blade or crankshaft damage. A walk-behind mower is less expensive and easier to maneuver than a tractor mower, but the latter type saves time and muscle power. If you plan to use a large mower, space your rows far enough apart so the trees won't be damaged during mowing even after they have grown for a few years. If, after mowing, you want to eliminate all weed competition and have your planting look especially neat, use a shoulder-carried string mower to trim close to the trunks where the wheel mowers can't safely reach.

CHAPTER NINE

Improving Wild
Native Trees

E very section of the country has areas where evergreens suitable for Christmas trees grow wild. We and many of our neighbors first started in the Christmas tree business by improving the wild spruces and balsams that were already growing on our farms. In fact, at that time I recall thinking that it seemed ridiculous to plant new trees when acres of firs and spruce were already growing in the neglected pastures.

Creating good Christmas trees from wild specimens proved to be a far bigger task than I had anticipated, however. Even though I didn't need to plant new trees, it took weeks to remove all the unwanted species and thin the potentially good trees so they would have room to grow. But raising wild trees can be profitable, so even after we started growing trees in a plantation, we continued to improve a small number of wild trees, mostly because nature is so generous in providing them. Only if you have a large

number of wild evergreens that are less than 8 feet tall, is it likely to be worthwhile to try to transform them into a profitable plantation. Improving such trees can also be an excellent way to acquire experience, whether you want only a few for your own use or a great many to sell. You can get income from trees that are already growing far more quickly than from planted seedlings, and you can even begin to cut and sell greens from culled trees the first season.

The first step in managing a wild tree lot is to remove all the species you can never sell as Christmas trees. This can be done anytime the snow isn't too deep to cut them off at ground level, but the best time is midsummer. Trees are less apt to resprout from the roots when cut at that time. Leave the cut trees on the ground among the good trees to serve as a protective mulch for the new seedlings sprouting on the ground. Take down large trees carefully so you don't damage the little ones, and cut them into smaller pieces so they will rot faster.

Uncultivated land is likely to be full of tree seeds and roots, which will continue to produce a supply of unwanted trees and shrubs for many years. Removing nearby, large weed trees will diminish seed production, but cutting out the persistent hardwoods often seems like a losing battle. One option is to spray Roundup, a systemic herbicide that kills the entire plant, roots and all. Roundup has been tested for many years and is rated as one of the safest herbicides in use because it is biodegradable and leaves no residue in the soil. Woody deciduous plants sprayed with it in late summer are usually completely dead by the following spring. Fortunately, evergreens have begun to go dormant by then so are quite tolerant of Roundup, and the spray, if carefully directed, will not damage them.

After you have eliminated the shrubs and trees of unwanted species, cut out any trees of your chosen species that will clearly never grow into good salable specimens: for example, those that are tall and have few branches, those that are one-sided and crooked, and any others with defects that can't be corrected by pruning and shearing. But don't sacrifice a tree merely because it has a scruffy base. By pruning off the lower branches you can coax it to grow into a nicely shaped tree a few feet higher on the stump.

When all the weeding and culling have been done, you will probably find that the potentially worthwhile trees remaining are too thickly spaced, and they, too, will need thinning. Trees need space to grow and sunlight to thrive, so thin them enough that no two trees of about the same age are closer than 5 feet apart. By leaving small seedlings of different sizes between the ones that you will harvest soon, you can be assured of continuous replacement without replanting. This thinning can be done any time, but if you do it in late fall or early winter you may be able to salvage the prunings and market them as greens, or make them into wreaths and other decorations.

To be sure of regeneration in your wild plantation, save a few large trees to produce seed. Choose the best strains for your "seed trees" and leave three or four per acre. Cut off the lower branches to allow light to get to the nearby seedlings, and to force the trees to produce seed earlier. Although the roots and seeds of weed trees in the soil will most likely continue to sprout for several years, gradually you can convert the area to a high percentage of the species you like. Future weeding chores will never be completely eliminated, however, because birds, squirrels, and even the wind will continue to plant seeds of other trees to assist nature in her battle against clear stands of any one plant.

As you cull and thin, cut out spaces for roads, unless the area is tiny. Besides making it easier to harvest the trees, roads also make it handy to inspect the trees frequently, and roads offer good fire protection.

Nitrogen Shortages in Wild Plantations

When I first "rescued" the over-crowded balsam firs from among the thick mixture of cedar and spruce on our pastureland, I expected that the liberated trees would grow rapidly and flourish, just as thinned out turnips and carrots do in the garden. Instead, the following year most of the trees turned yellow and made little growth. I finally realized that the soil was filled with old roots and that the bacteria needed to decompose the roots that were grabbing all the nitrogen from the soil. The poor trees were struggling along with what little fertility was left.

I remembered how we always used nitrogen fertilizer to speed up the rotting of old tree stumps in lawns so I began applying a high-nitrogen fertilizer around the trees. After both the trees and the rotting roots received the nutrients they needed, the trees began to grow and quickly regained a rich green color.

To avoid the same problem I had, scatter a high-nitrogen fertilizer such as 15-10-10 around the area the same

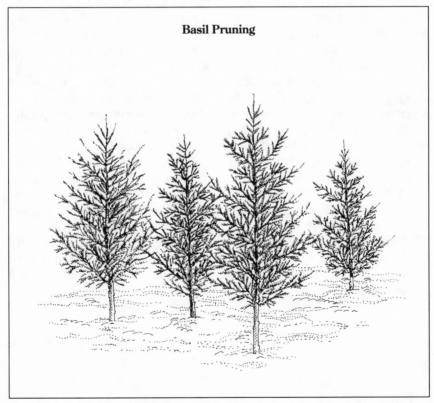

Basil Pruning

To cultivate wild trees, prune away all of the lower, straggly branches below the first complete whorl.

year you liberate the trees. Continue to feed them each spring, because conifer roots rot slowly and those of large cedar trees may remain in the soil for a decade or more. Trees under stress from a nutrient shortage not only look sick and grow poorly, but are also less able to resist the extremes of weather and attacks from disease and insects.

Pruning and Shearing

After culling, thinning, and fertilizing, next on the program in wild tree management is to prune off the straggly lower branches of the taller evergreens up as far as the first complete whorl (circle of branches around the tree). Removing these branches forces energy into the upper ones, and causes them to grow thicker and faster. Opening up the base also allows more room and light for the little replacement seedlings growing beneath the tree. Late summer or fall, when the tree is dormant and doesn't bleed, is the best time for this chore. You can do it more rapidly with an axe, but clippers will do a neater job, with less damage to the trunk of the tree. Cut close to the trunk so no stub is left. Shearing is the final touch that turns what might otherwise have been a wild, junky tree into a tight-growing, premium specimen. Shearing the

tall evergreens that you are likely to have in a new wild area can be slow and tedious, however. Pole pruners are a necessity, and those with lightweight aluminum extensions are far easier to maneuver than the ones atop heavy wooden poles. When we have a lot of trees to shear, I like to alternate between shearing the tall and short ones, both for the sake of variety and to create the illusion of saving human energy. (For more information about pruning and shearing, see pages 73-78.)

Even with seed trees producing offspring in the plantation, some replanting may be necessary. Firs and spruces usually reproduce very well from natural seeding, but pines tend to reseed far less vigorously, so most wild pine stands will need your help. Provide any necessary replacements with nursery-grown stock or transplant native seedlings from recently cut-over areas where they usually grow abundantly.

In spite of the labor involved, wild plantings have many advantages. Not the least of them is that since trees of all sizes are growing together, many more trees can be produced in the same area during a 10-year period. By contrast, in a field of well-spaced, planted trees, a lot of room is wasted during the first few years.

II
GROWING PREMIUM TREES

Growers who do only a half-hearted job of caring for their lots are certain to end up with mostly run-of-the-mill, inferior Christmas trees. There is no longer a market for these low-quality trees, and the supply of high-quality ones is certain to increase in the future as the industry becomes increasingly competitive. It is important, therefore, to produce only the best, and I hope that this section of the book will help you to do just that. By maintaining good soil fertility, shearing the trees properly, protecting them from pests, and managing your lot well, you will be able to produce a crop of blue-ribbon trees every year for as long as you decide to stay in business.

CHAPTER TEN

Maintaining
Soil Fertility

J ust as good nutrition and a wholesome environment produce healthy people, soil fertility is the key to healthy plants. Trees that are well nourished not only grow faster, but they have better color, more resistance to disease and insects, and the ability to cope more successfully with drought, cold, and other weather extremes. Of all the commonly grown Christmas trees, Scotch pines are the least fussy. It is possible to grow them on soil that is less than Grade A, but they will respond better when they have adequate nourishment.

If your soil was in prime condition at planting time, as described in Chapter 8 (see especially pages 55), little maintenance will be necessary, because trees don't deplete soils as do corn or potatoes. Like all crops, however, they require certain nutrients and these must be available in a form that the trees can absorb readily.

Scientists tell us that twenty different elements are found in the tissue of a healthy tree and all are needed for good growth. They include carbon, hydrogen, and oxygen elements that make up the organic compounds found in all plant and animal life, and are derived from air and water. Trees also need nitrogen, phosphorus, potassium, calcium, and magnesium, all familiar to farmers and gardeners who buy fertilizer. Necessary also, but in much smaller amounts, are trace elements, including boron, chlorine, cobalt, copper, iron, manganese, molybdenum, silicon, sodium, sulfur, vanadium, and zinc. In much of North America, wherever soil is in good condition, enough of each of these trace elements is available to supply the minute amounts needed by Christmas trees. Some regions, however, have soils with one or more of them lacking. Boron, zinc, and manganese, particularly, are often in short supply and must be supplemented. The extension soil service (or, in Canada, Agriculture Canada) may be able to warn you about local deficiencies and may recommend a laboratory analysis of your trees' foliage.

The fertilizers commonly used by Christmas tree growers are granular types sold in either 50- or 80-pound bags or delivered in bulk and spread by fertilizer dealers. Since trees need nitrogen especially, most growers prefer the ones with high nitrogen formulas, such as 10-5-5 or 15-10-10, rather than the 5-10-10 types used by gardeners. (The numbers refer to the percentages of the most needed nutrients — nitrogen, the first number, phosphorus the second, and potassium the third. The remaining percentage is an inert filler.) If a soil test indicates a need for magnesium or some other trace element, a fertilizer that includes it, represented by a fourth number (for example, 10-5-5-2), may be used. For large growers who use huge amounts, fertilizer companies will supply formulas especially mixed according to whatever an analysis shows that the soil requires. This saves growers the expense of buying and applying nutrients that their soil doesn't need.

Organic fertilizers can also be used on Christmas trees, but seldom are, except in small plantings. Because they release nutrients slowly, they are less apt to burn the tree, but this fertility is also more likely to be absorbed by the sod cover, and not reach the tree roots at all. Also, nitrogen, the element most needed for good tree growth and the one lacking in most soils, becomes very expensive when spread in the form of manure, cottonseed, soybean, blood meal, or other natural sources.

Fertilizing Methods

How to best feed a growing tree can work up a lively discussion at a meeting of Christmas tree growers. Many people dump a small amount of 10-5-5 in one spot on top of the soil just outside the branch spread of each tree in early spring, increasing the amount each year as the tree grows larger. They maintain that the fertilizer feeds the tree even if it reaches only a few roots, and they like the fact that the large amount of nitrogen placed in one spot kills the grass, so it doesn't compete with the tree for the lunch. Other growers feed a tree by cutting a slit in the soil with a shovel and dropping in a cup of fertilizer, because they believe this encourages deeper rooting. They are concerned that in dry years, fertilizer placed on top does not move downward rapidly enough to be useful.

I like another method, one endorsed by many growers, especially those who keep their plantations in sod. In early spring, fertilizer is spread completely around the tree, just outside the perimeter of the branch spread. Growers who favor this technique theorize that when the entire soil area around the tree is well fertilized, more roots can absorb the nutrients, and the tree will grow better.

Whichever method you choose, always place the fertilizer just outside the spread of the branches so the roots will reach out to it. Placing the nutrients close to the trunk encourages shallow rooting, with the unhappy result that the tree is less able to locate the trace elements it needs and is less able to find moisture dur-

Feeding Fertilizer to a Tree

A good method of fertilizing is to spread fertilizer completely around the tree, just outside the perimeter of the branch spread, in early spring.

ing dry spells. Fertilizers can be carried in a pail and applied by hand, but with a back-carried spreader that releases a measured amount of the plant food through a tube, you can apply just the right amount to each tree, with far greater ease.

With fertilizer, as with all chemicals, more is not better. In an effort to get the fastest growth from their trees, however, many growers overfeed them. Using excessive amounts of chemical fertilizers not only wastes money but is likely to "burn" the trees by drawing water away from the roots. This will show up by a browning of the needles, and can be extreme enough to kill the tree. Make sure, too, that the fertilizer is applied directly to the ground and not on the branches of the tree, where it can burn the foliage by direct contact.

Good color and steady, rather than excessively fast, tree growth should be the goal of your feeding program. For one thing, trees that grow too fast need additional shearing to keep them bushy. In addition, top sprouts that grow abnormally fast are weak and easily broken off in high winds or even by resting birds. Fast-growing trees are also more subject to disease and to aphid and other insect attacks. Finally, overfeeding, like undernourishment, may result in winter injury and other forms of stress from extremes in weather.

Since guessing the right amount of fertilizer to use is not easy, some growers use slow-acting fertilizers such as Mag-Amp. Because their nutrients are released slowly over a long period of time, damage to the trees is less likely even if too much is used. Slow-release fertilizers also give the grower more leeway in application time, because the fertilizer can be spread in early spring when there is less to do in a plantation and still be available when the trees are ready to use it.

To Lime or Not to Lime....

If your soil has been tested and the correct amount of lime applied for the species of trees you are raising, it is unlikely that you will need to add more during the life of the crop. Exceptions could include very light soils where the calcium leaches out rapidly or those rich in aluminum.

As I said before, if you are raising pine trees, only soils with a pH of 5.0 or less are likely to need lime. Firs and spruce prefer less acidity and are more likely to benefit from lime, but test your soil first to be sure of the need. Too much lime has the same effect as too little—it locks up the soil nutrients so trees cannot utilize them. One hundred pounds of lime spread over 1,000 square feet will raise the pH there one point.

Unlike fertilizer, lime can be put on at any time of the year, but it is better not to apply it at the same time as chemical fertilizer. Sometimes an interaction between lime and certain types of fertilizer makes the latter less effective. Allow at least one good rain between the applications, if possible.

Soils in North America are seldom too alkaline, unless they have been heavily limed recently, and if this is the case, the condition will usually correct itself within a few years. An exception may be in seed and transplant beds where overly alkaline soils encourage damping-off diseases that kill seedlings. Soil can be made more acid by spreading sulfur at the rate of 3 to 4 pounds per 100 square feet to lower the pH one point. Mixing peat moss into the soil will also make the soil more acidic, as will using a fertilizer with an acidic base, such as ammonium sulfate, rather than the more commonly used ammonium nitrate.

CHAPTER ELEVEN

Shearing and Pruning

Whenever the teacher at the rural, one-room school of my boyhood needed a small Christmas tree for a play, we always looked for one that the cows or deer had nibbled on. These often grew bushy and occasionally had been trimmed into a perfect shape. With so many obvious exhibits standing in farm pastures, I have no idea why it took growers so long to imitate the cows and shear their Christmas trees.

Wild trees are more anxious to grow large than to become holiday decorations, so few ever get dense without help. Spruces are naturally thick-growing and, even when uncared for, they occasionally develop into good Christmas trees, but the sparsely branched firs seldom do, and the pines, almost never.

Early Christmas tree growers working with wild trees discovered that by cutting off the bottom branches they could stimulate the tree to grow more densely for the next few years; and for a decade or so,

this was about all the pruning trees got. The tops of these basal-trimmed trees always remained thin, however, and a lot of years were wasted while growers let a tree grow 6 feet tall before cutting off its lower 2 feet of branches and then waited several more years for it to reach salable size.

Basal pruning is still practiced to get a nice stem on the tree just below a good whorl of branches, but growers learned to shear by duplicating the methods used by professional landscapers, many of whom had come from Europe where gardeners had shaped trees in formal gardens for centuries. One early shearing pioneer was nurseryman John Young, who started a Christmas tree plantation in northern New England in the 1960s. His trees soon became the standard for the industry, and growers from all over the eastern United States and Canada searched out his high mountain farm to study his techniques.

The terms pruning and shearing are often used interchangeably, but

professionally speaking, pruning refers to the cutting off of woody por-
tions of a tree, and shearing, to the
cutting back of soft new foliar growth.
Therefore, cutting off the lower
branches or removing an extra top
that is more than a season old is
considered pruning, but clipping away
the soft new needles, as in trimming
a hedge or topiary shrub, is regarded
as shearing.

Shearing and Pruning Tools and Machines

Different tools are used for the two
cutting jobs. The most widely used
pruning tools are the following: hand
pruners, used for both light pruning
and early shaping of small trees; long-
handled pruners (loppers), used for
heavier cuts; pole pruners (pruners
at the top end of an aluminum or
wood pole that are worked by pulling
a cord), good for pruning or shearing
tall trees; and saws, both hand- and
power-operated, used for removing
cull trees, cutting off limbs that are
too big for pruners to handle, and for
that final pruning task, harvesting.

Shearing tools include thin,
machete-like knives, power clippers,
and hedge shears. Most shearing
tools are available with either medium or long handles. I prefer the
long-handled types because I can
reach further up on tall trees with
them as well as shape smaller ones
without bending over. The shearing
knife is the most popular hand-shear-
ing tool for Christmas tree growers
because it does a fast, efficient job
and is easy to sharpen. It must be kept

very sharp to work well, so it is wise to
wear a leg guard and heavy gloves
both when shearing with it and when
sharpening it. I always carry a sharpening steel or stone in my left hand
whenever I am using the knife, because it prevents me from ab-
sentmindedly reaching out in front of
the swinging blade.

Large growers use gasoline-
powered shearing equipment with
reciprocal blades that resemble electric hedge clippers. These trimmers
come in lengths of up to 6 feet and are
very efficient to operate. Most are
hand held, with the engine in the
handle, but some are powered by an
engine carried on the back.

Pruning Techniques

Sometimes it is necessary to cut away
an extra top or two, or drastically to
shorten a branch that is out of bal-
ance, but the chief pruning procedure used by most Christmas tree
growers is to cut off the lower
branches of a tree when it is about 4
feet tall. Basal pruning, as it is called,
involves the removal of all the thin
lower branches up to the first good
whorl of branches. Some recommend
1 to 1 1/2 inches of bare stem for each
foot of tree height. This procedure
not only ensures evenly spaced
branching at the bottom of the tree,
but provides an open stem near the
ground that makes it easy to saw off
the tree at harvest time. Always make
basal pruning cuts close to the trunk.
Stubs with no green needles growing
on them will die and may rot and form
open wounds or cankers. (See illustration.)

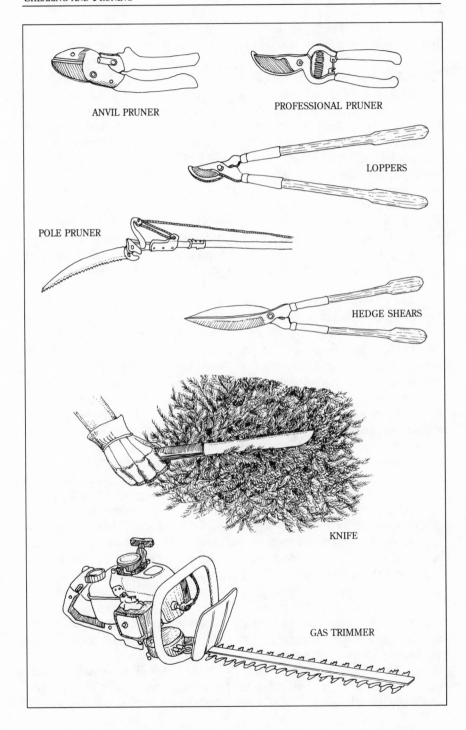

ANVIL PRUNER

PROFESSIONAL PRUNER

LOPPERS

POLE PRUNER

HEDGE SHEARS

KNIFE

GAS TRIMMER

Shearing Techniques

It is a mistake to delay shearing until the tree is 4 or 5 feet tall. At that stage it is difficult to coax it to "tighten up," and even if it does get more bushy on the outside, it is likely to look hollow in the interior. Instead, start to shear lightly when the tree is about 2 feet tall; but let it grow gradually larger. Each year heavier shearing will be necessary, to get the tree into the tight form and tree shape you want. Pines grow more rapidly and require annual shearing, but unless a very dense tree is wanted, firs and spruces may need shearing only once every two years.

Trees are programmed to grow, and shearing, when done at the right time, makes the best use of that instinct. Shearing checks the natural upward and outward growth of the tree and the tree responds by sprouting new buds between existing branches, causing growth where it would not otherwise have taken place. Since the growing season of most conifers lasts only three or four weeks, the optimum shearing time is short, and it is important to begin in late spring or early summer, as soon as the new sprouts are two or three inches long. Large growers, especially, must plan carefully and work fast to get the job completed on all their trees before new terminal buds set. Shearing during the growing period allows enough time for the redirected growth to take place before the season ends and it also enables the cut ends to heal over and form new terminal buds that will sprout and grow the following year. Trees sheared at the right time heal so effectively there is little evidence that any shearing has taken place.

Pines are nearly impossible to force into a good shape if they are pruned at the wrong time. Spruces and firs can be sheared at other times of the year, but tend to have irregular tip growth and form extra tops that must be removed.

Whenever lots of shearing must be done quickly, there are many models of power machinery available, although hedge shears and a shearing knife may be adequate for small plantations. Lightweight, gasoline-powered clippers come with cutting bars in various lengths up to 8 feet long. Long cutter bars permit a fast trim of an entire side of a tree in one sweep, but the shorter ones require less skill and muscle power.

Some growers prefer electric clippers because they are quieter, vibrate less, and require little maintenance. In a large plantation, these clippers are usually powered by a tractor-operated generator, which can be moved about the plantation easily and supply current for several clippers. Electric clippers do have drawbacks, however. They are dangerous to use in wet weather, and not only must the cords be continually dragged around, but they easily become entangled in the trees. I have on more than a few occasions accidentally clipped off the cord amid a shower of sparks that invariably preceded a time-consuming splicing job.

Cordless clippers may be suitable for a very small planting, especially since models with longer-last-ing batteries have recently been developed.

It will help you to become a good shearer if you work with, or at least first observe a professional at work. Shaping a tree is difficult to learn from a book, but it will help always to keep the normal shape of the species in mind. Spruces and firs grow naturally into the shape of a slender pyramid, much like an inverted ice-cream cone. Pines are more spreading, so a slight, archlike bulge at the middle looks natural, but be careful not to create a house-roof effect with sloping sides on the top half of the tree and near-vertical sides on the lower half.

Odd shapes can best be prevented by shearing in one long sweep from top to bottom as you go around the tree, instead of first shearing the top half and then the bottom. This technique is more efficient, too, because by circling each tree only once, a lot of time and mileage is saved during the course of a day. A skilled shearer, using a shearing knife, can usually do sixty or more medium-sized trees per hour.

Although a premium tree should be a tight specimen, your goal should *not* be to produce one as thick as possible. When sheared too tightly, trees tend to develop a thick stem that adds considerably to the tree's weight and thus to the difficulty of transporting it and mounting it in a plant stand. Too much density in a pine traps a huge amount of old, dead needles in the interior that are difficult to shake out when they are damp; but they are almost certain to dry out later and

then drop on the living room rug, greatly tempting the customer to shop for an artificial tree the following year.

Tastes in trees change as in other fashions, and shaping methods must be adapted to what the public wants. A few decades ago the judges who graded trees had different standards of what constituted a premium tree, and probably most of their blue- ribbon choices then would not get third place today. Lighted candles, which were popular during the early part of this century, required a loosely branched tree, but electric lights make no such demands. Wide, flaring trees were the style when growers first started to shear, but they were too wide for many small apartments and mobile homes, as well as houses without large living rooms. Now a tree with a narrow taper is more in demand. Growers who have been slow to change their shearing methods have consequently often been stuck with a lot of fat, unwanted trees.

Tastes in trees vary also from place to place. A grower should investigate what is preferred in his marketplace before deciding how to shear. Those who sell retail or operate a choose-and-cut farm need to offer a variety of shapes and sizes so their customers will have a good choice, but wholesalers usually prefer loads of trees that are fairly uniform in size and shape.

After the shearing job is completed, take frequent walks through the plantation throughout the summer with knife or pruners in hand. Late growth, such as an extra top or

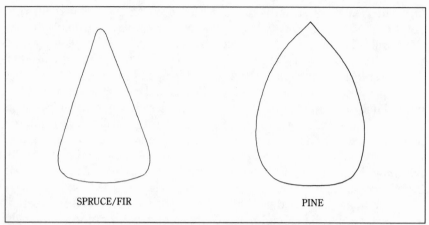

Prune pine trees to a slight archlike bulge in the middle, according to their naturally spreading shape; prune firs and spruces to their slender, pyramidal shape.

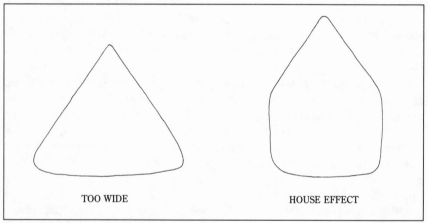

Avoid too-wide or house-effect shapes by shearing in one long sweep from top to bottom as you go around the tree.

branch in the wrong place, should be removed before the growth hardens. After a few years of shearing, you will be keenly aware of the consequences of each cut and be able to shear rapidly and easily. I hope you'll enjoy it and find it a satisfying job. As the great horticulturist Liberty Hyde Bailey once said, "Of all the operations connected with horticulture, pruning, shaping and training bring the person into closest contact and sympathy with his plant."

The Fine Art of Weed Control

Most of the early Christmas trees were raised on farms, and my neighbors solved the grass and weed problem by simply letting their cows and horses pasture in their plantings. This worked well if they remembered to keep the animals out of the tree lot during early summer when the tender new growth proved tempting to the livestock and the grazing deformed the trees.

Pasturing of animals is still practiced on some tree farms, especially where wild trees are being raised. I know of one grower who allows his chickens and geese to roam in his small tree lot all summer. It is tricky, however, to allow just the right amount of grazing to control weeds and hardwood brush, yet not damage the trees. In any case, sheep and goats, which browse woody plants heavily, should never be used for weed control because they are too destructive to be allowed with trees.

Trees on some plantations are not grown in sod but in bare earth. In these lots, grass, weeds, and woody deciduous brush are usually kept out entirely by chemical herbicides, but occasionally the soil is cultivated with a rotary power tiller. When trees are grown in sod, however, mowing (as described on pages 61-62) and herbicides are the usual methods of weed suppression.

Herbicides

When weed killers were first developed, they were hailed as a great agricultural triumph, and many expected them to solve the weed problem forever. Like DDT and other chemical breakthroughs, however, some have created more problems than they solved, and many of the early herbicides were found to cause cancer and contaminate water supplies. As a result, their use is now restricted, and a few have been taken off the market.

Fortunately, recently developed herbicides have been more thoroughly tested before being intro-

Some Herbicides Commonly Used in Christmas Tree Culture

Trade Name	Active Ingredient	Manufacturer
Aatrax	atrazine	Cieba-Geigy
Devirnol	napropamide	Stauffer Chemical
Fusilade	fluazifop-butyl	ICI Americas, Inc.
Garlon	triclopyr	Dow Chemical
Goal	oxyfluorfen	Rohm and Haas
Gramoxone	paraquat	ICI Americas, Inc.
Kerb	pronamide	Rohm and Haas
Poast	sethoxydim	BASF - Wyandotte
Princeps	simazine	Ciba - Geigy
Roundup	glyphosate	Monsanto

duced, and some sprays deactivate soon after they start the demise of the weeds. Roundup, one of these, is extremely popular and commonly recommended because of its effectiveness and reputation for being safe. Although I do not recommend depending on chemicals in a small planting where other methods are more practical and ecological, I do feel that herbicides have their place when the safest ones are chosen and used carefully. Some of those most often recommended for tree culture are listed and described in this chapter.

Even though the new herbicides are touted as safe, when you decide to use any chemical, do so with care. Follow the directions exactly, such as wearing protective clothing when so advised and never applying more than is recommended. Clean the sprayer thoroughly immediately after using, and don't use it for any other purposes, such as the application of insecticides, fungicides, or tree colorants.

If you "spot spray" rather than cover the entire area, you may want to add a small amount of paint to the spray mixture so you can see where you have sprayed. Use water-soluble paint for sprays that will be dissolved in water, as most are, and oil-based paint if the herbicide is to be mixed with oil or kerosene.

Keep accurate records of each chemical purchase you make, when and where you apply it, and the results obtained. Do not buy more than one year's supply at a time, because some herbicides lose effectiveness in storage and, as many growers have found to their dismay, new information may be forthcoming that the product has been deemed unsafe.

Herbicides vary greatly as to what they accomplish and how they do it, so study the information supplied by your extension service, Christmas tree association, and companies that manufacture them before you decide which ones best fit your needs. Some injure foliage, so the spray must be directed away from the tree's needles; others are not safe to use on evergreens during the growing season, but can be safely applied in late summer or early fall after new growth has hardened. In addition, pines, firs, and spruces all vary in their susceptibility to damage from the various herbicides.

Don't be surprised if the results from spraying vary considerably each time, even when you carefully follow the directions. The effectiveness of some chemicals depends on the type of soil, the time of year, the air temperature, subsequent rainfall, the kinds of weeds and their growth stage, and other factors. By keeping records, you will become more knowledgeable about the best way to use an herbicide and get consistently good results from it. By the time you have this all figured out, though, your chosen chemical may be replaced with another one, and you will have to start over again!

How Weed Killers Work

Herbicides come in various forms. There are liquids and powders that can be mixed with water before application, and granular kinds that are sprinkled on the soil like fertilizer. Selective herbicides, which kill only certain weeds or grasses, are used on lawns, golf courses, and farm crops such as corn. Nonselective herbicides, which destroy all vegetation, are used around trees, on driveways, under fences, on rights-of-way, and sometimes on fields to kill grass and

weeds before planting a crop. Both selective and nonselective herbicides work in one of the following ways:

Foliar. The plant is killed when a chemical is sprayed on the foliage and absorbed by the plant. Among those in this category are 2-4D, the well-known broadleaf weed killer; Roundup, which is effective on most growing plants; and Fusilade, a grass killer. Some foliar-type herbicides can also be used to spray on hardwood stumps to keep them from resprouting.

Soil Activated. Herbicides such as simazine and atrazine are sprinkled or sprayed on the soil to be absorbed through the roots of weeds, killing them. Some, like simazine, are formulated to kill established weeds, but others, such as napropamide, act as preemergents, preventing the sprouting of dormant seeds, including any that blow in during the period the chemical is still active. To be effective, most of these soil-activated herbicides must be incorporated into the soil by application just before a rain or irrigation, or by tillage. Preemergents remain active for only a few months, and when applied as directed, do not kill established plants.

Restricted Chemicals

Many herbicides, as well as certain insecticides and fungicides, that were once commonly used by Christmas tree growers and other farmers are now regulated by federal and state laws. The government wisely wants to keep track of the amount being used, and where. Consequently, in

Cautions for Using Agricultural Chemicals

ALWAYS read the label and instruction folder closely before buying to make sure that the product will meet your needs and be safe to use.

ALWAYS read the label and directions each time before using. Follow all recommendations.

WEAR protective clothing when recommended.

NEVER spray at a higher rate than recommended. Measure all ingredients carefully; never guess.

DON'T use in a way so the wind will blow spray toward you.

WASH sprayers and other containers after use.

WASH thoroughly or dispose of any contaminated clothing and bathe immediately if any skin has been exposed to chemicals.

STORE all chemicals in a locked cabinet, in an building that is not likely to be ignited in case of a fire.

DON'T buy more chemicals than you can use in a year or two. Dispose of any that are of uncertain age.

order to buy and use these restricted chemicals, even on your own property, you must obtain a permit. Several hours of instruction and a written test are necessary to get this license and it must be renewed periodically.

Your county agent or forester (in Canada, contact Agriculture Canada) can give you an up-to-date list of which chemicals are restricted and tell you when instruction classes are being held.

CHAPTER THIRTEEN

Things Can
Go Wrong

Any enterprise that involves living things can be full of surprises. I don't agree with the popular wisdom that you must spend fifteen years with each farm project to be sure of making every possible mistake. We have been growing trees much longer than that and not a year passes without new challenges. When I began to grow Christmas trees, I already had years of experience growing vegetables, fruit, and nursery stock, and had long since accepted the fact that coping with weather, insects, disease, and wild animals was a part of managing these projects. Evergreens, though, would not have problems, I expected. As far as I knew, they just grew and nothing bothered them. Forest trees died only when they were overcrowded, cut down, or reached old age.

I soon found out how wrong I was, however, and that conifers suffer from a wide range of pests, too. Native species were not immune, and those brought in from other regions seemed to have even more problems. This chapter is meant, not to be pessimistic, but to help prepare you for any problems ahead and, it is hoped, to help you avoid as many of them as possible.

Weather Eccentricities

Bad weather is hard to predict and even harder to control. Some years, conditions are perfect for growing, with ideal amounts of rainfall each week — and those years we treasure. But weather isn't always made to order, even if you grow trees in the best possible climate. Unfortunately, extended droughts can wipe out entire new plantings, and even people who can afford to irrigate may not find enough water available. Summer hailstorms and high winds are usually less damaging to Christmas trees than to squash vines and peaches, but severe storms may snap off tender top leaders, spoiling the future growth of the trees.

Winter cold seldom bothers conifer species that are suitable for the area, but even the hardiest may sustain damage from ice storms. Pines, especially, can be badly damaged when a hard frozen crust forms on top of deep snow, then settles as the snow melts from beneath and rips limbs from the trunk as it sinks. Alternate freezing and thawing in years with little or no snow cover can heave young trees out of the ground, drying out their roots.

Winter-burn—browning of needles—occurs when winter winds evaporate water from the trees' needles faster than it can be replaced by the roots, especially when the ground is frozen for long periods or when the trees are growing on poor soil that tends to be dry much of the time. It can affect any evergreen, but some species are more susceptible than others, and it is more likely to occur on young trees with shallow roots. Minor damage disappears as it becomes covered with new growth, but severe windburn may deform or even kill a tree. To minimize this and other winter damage, plant your trees in good soil and keep them well fertilized. Trees in a plantation that is maintained in sod rather than bare soil are less likely to become winter-burned, because the grass thatch serves as a protective mulch against the cold.

Along with winter-burn, another culprit is salt. Once it was only trees planted near the seashore that were damaged, but now those growing close to highways where traffic kicks up road salt in winter can suffer, also. The fact that wind carries this salty spray for considerable distances is strikingly obvious along heavily traveled northern highways every spring. Evergreens can also be damaged when salt-laden snow melts and drains onto their roots. These conditions are difficult to control in vulnerable areas, so it is wise not to plant trees where salt damage is likely.

Spring doesn't end the problems, either. Late spring frosts are a common complaint of tree growers nearly everywhere. Typically, after a few warm days have stimulated the trees to start growing, a freezing night comes along and all the tender new growth is killed. The trees usually resprout, but the second growth is uneven, and major shearing is necessary to get the tree back in shape. It helps to avoid planting in pockets where cold air is likely to settle, but your best insurance is to plant only species that are acclimated to conditions similar to yours so they won't start growing before warm temperatures arrive to stay.

Plant Stress

Like humans, plants suffer from stress, and weather extremes make them more susceptible to attacks by insects and disease. Trees sometimes appear to have survived flood, drought, or cold, and then succumb a few weeks later. They have simply become too weakened to absorb the needed nutrients and convert them to plant tissue.

Stress can be caused by circumstances other than weather, of course.

Environmental factors such as air pollution or a heavy rainfall weaken trees, as does too little sunlight or too much competition from weeds. Certain soil conditions also contribute to stress: excessive wetness, acidity, or alkalinity; deficiency in a necessary trace element; or an excess of aluminum. Damage to the bark caused by deer, rabbits, mice, or mowers can be devastating as well. Stress can also be caused by a grower who applies too much or too little fertilizer, uses herbicides or insecticides improperly, or prunes off mature branches during the spring when the trees tend to bleed badly.

Natural Controls of Diseases and Insects

Whenever a forest consists of several species of native trees, it is seldom seriously bothered by diseases and insects, but whenever one species dominates, nature takes steps to correct the unnatural situation and tries to restore a more balanced mixture of plant life. Every few years tent caterpillars viciously attack the chokecherry hedgerows that grow alongside the backroads in our town, for example, and for the next few years the less vigorous shrubs have a chance to compete. In the same way, blights and insects attack orchards, fields of potatoes, and wheat. In a mixture of plants, not only is it harder for a disease or insect to find the next plant of the same species, but the wide variety of plants supports an assortment of bird life and predatory insects that helps keep destructive insects in check.

Naturally, we Christmas tree growers don't relish the idea of allowing unprofitable plant life to thrive on our valuable land, but there are ways to create variety without losing much growing space. If you grow more than one species of Christmas tree, alternate small plots of different species, rather than planting all those of the same species in a large block. When you are preparing a field for planting, don't be too hasty about clearing out all nearby tall trees that can serve as nesting places for hawks and other birds. Even tall dead trees can be useful. They may not add beauty to the landscape, but they make good apartment houses for owls, woodpeckers, and other hungry birds.

By dividing plantings with hedgerows of deciduous shrubs some planting area is wasted, but the practice may greatly lessen the need for spraying. When planting such hedgerows, choose the least weedy kinds of plant life such as viburnum, wild apple, wild plum, and mountain ash. They are good bird attractors and will not spread as vigorously as willow, poplar, wild cherry, and alder.

You can avoid some unpleasant invasions of insects and diseases by ordering trees only from reliable nurseries. Many of the most serious tree diseases, such as scleroderris canker, have been spread throughout the country by the careless shipping of infected seedlings. Your extension service and Christmas tree association can help you find reliable nurseries, and aid in choosing the species that are most resistant to the diseases and insects that may be prevalent in your area.

You can avoid many stressful conditions by keeping the soil in healthy condition and by using chemical fertilizers and herbicides only with the greatest care. But even when you use the best cultural methods and preventative measures, a disease or insect menace can appear, so in order to save your investment, be ready to act fast.

Taking Action

It may be tempting to reach for the spray can immediately at the first sign of a sickly looking plant, but don't load up the sprayer unless you know for sure what's wrong. First, take time to determine what is causing those dead needles or drooping new growth. Many tree disorders are physiological, and rather than indicating the presence of a serious disease or insect pest, the condition could be a symptom of stress caused by any of the factors described previously. When you are certain it is not an environmental problem, look for tiny eggs or larvae of insects, or the fruiting bodies of disease (spores, rust, molds), to help decide on the course of action.

To help in your identification, invest in a hand lens that you can carry in your pocket, or if you want to look like a professional, wear it around your neck. Also indispensable is "The Christmas Tree Pest Manual," available from the United States Department of Agriculture Forestry Service, Superintendent of Documents, U.S. Government Printing Office, Washington, DC 20402. It has excel-

lent color pictures that will help you identify tree insects and diseases.

If you are unable to identify a disease or insect, your county forester (or Agriculture Canada) should be able to help, or if not, tell you how to contact the state plant pathologist. Don't hesitate to call these officials, because they want to keep track of possible epidemics. Although a few new pests appear from time to time, most of those you will encounter have always frequented forests, but seldom caused any alarm. Few ever damaged the valuable wood, and until people began growing Christmas trees, no one worried about the ugly appearance of a few needles.

Even though the potential danger of a disease or insect epidemic can be very real, don't panic every time something looks slightly unusual. Each year a few frightened tree farmers call the extension service to report dead and dying needles throughout the interiors of their trees. They have forgotten that, unlike deciduous trees, evergreens do not drop all their foliage at one time, but instead lose part or all of the older needles at sometime during the year. Some species annually dump all the needles that are over a year old and others shed only those that are two or more years old.

Most diseases thrive in humid conditions, so are most troublesome during wet seasons. Many insects, on the other hand, appear to have the largest families and the shortest periods between generations when weather conditions are warm and dry. It can be a no-win situation, especially during those years that have extended periods of first one kind of weather and then the other.

I do not suggest any particular chemicals for controlling insects and diseases, because the recommended list of pesticides changes frequently. Old standbys are occasionally found unsafe, and safer and more effective chemicals are constantly being developed. Check with your state or county extension office (or Agriculture Canada) or farm store to see what treatment is currently being advised for your problem.

Read all pesticide labels carefully before buying them to be sure that the product is safe to use on trees and recommended for the pest affecting your planting. Some fungicides control only certain diseases, but not others, and insecticides that kill weevils are not always effective on mites. It goes without saying that an-

other reason why we should read all the directions on the label is to understand how to mix the spray solution properly and how much to apply. Pay close attention to every caution, not just for your own safety, but also for the sake of your trees. We know of one grower who killed a whole field of trees with a chemical that a salesman for his supplier wanted him to try. In reading the label later, he found that the man had given him the wrong information, but since it was all verbal, he couldn't prove a thing, and as a result lost thousands of dollars. Read the label and use the product only on the species that it has been registered for.

Sprayers and Spraying
Be especially careful in applying chemicals with toxic properties. Sprayers that produce a fine mist give the best coverage to trees and make

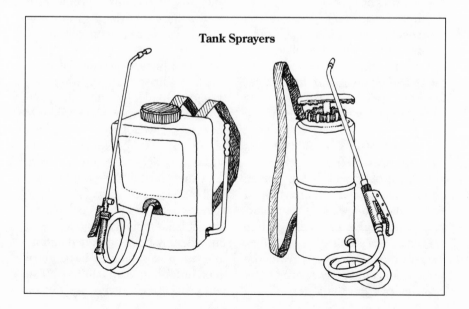

Tank Sprayers

the most effective use of the material, but are also the most likely to drift and collect on skin and clothing and be breathed into the lungs. Follow the recommended timing and precautions, spray when there is little or no wind, and always wear protective clothing when it is advised.

Diseases and insect problems usually pop up suddenly, leaving little time to shop around for equipment, so investigate what problems other growers in your area have experienced recently and have the necessary equipment and pesticides on hand. A heavy-duty, hand-carried tank sprayer can protect a lot of trees and requires far less maintenance than a power-operated type. For a larger operation, you will probably need a back-carried, gasoline-powered mist sprayer, however, and for the very large ones, a tractor-operated, high-pressure machine similar to those used by orchardists.

If you use herbicides, you will want two sprayers because it is not a good idea to use weed killers in a sprayer that is ever used for other purposes. Sprayers are hard to get perfectly clean, and even tiny amounts of some herbicides can injure a tree.

Common Christmas Tree Diseases

Brown Spot. Primarily hits Scotch pine, causing a reddish-browning of the needles, followed by their dropping. Fungicides can help, but to prevent its spread to other plantings, cut all infected trees close to the ground, including any nearby older Scotch pines.

Canker. Large patches of white pitch on the trunk and branches usually indicate canker presence. Canker wounds are caused by a fungus that infects trees, especially those stressed by winter injury, breakage, and bad pruning wounds. Although it can kill larger trees, it usually only disfigures those of Christmas-tree size. Infected branches should be cut back to the trunk as soon as you notice them. To help prevent cankers, make all basal pruning cuts close to the trunk and avoid injuring the bark. If the disease is present in the neighborhood, it is best not to shear trees during wet weather, because the spores spread more easily then. Fungicides help control existing cankers.

Cytospora canker affects spruces, primarily the Colorado and Norway species. *Diplodia canker* and *shoot blight* affect Austrian, red, and Scotch pines, often killing small seedlings. The tips of terminal sprouts turn brown, die, and curl up. An outbreak of *scleroderris canker* greatly upset the Northeast a few years ago, causing quarantines that resulted in large losses by Christmas tree growers. Although it may hit spruces, firs, and Douglas fir, it is a problem primarily on pines, particularly the red and Scotch species. The North American strain kills only trees that are less than 7 feet, but the European strain is fatal to trees of all ages. The disease is extremely difficult to control, and once it is identified, the area should immediately be quarantined and no trees sold. Fungicides have given good control in seed and transplant beds, but are often not effective on larger trees.

Diplodia tip blight and canker infects Austrian, red, and Scotch pines, killing the new shoots on trees of all ages, including nursery seedlings and transplants. The end shoots curl up and turn brown, and olive green streaks can be found beneath the bark. If the disease persists, the canker eventually girdles the tree and kills it. A fungicide sprayed every two weeks during the growing season helps prevent the disease from spreading.

Needlecasts. There are many so-called needlecast fungus diseases, which cause heavy dropping of needles before the trees are harvested. Most affect pines primarily, but sometimes other trees are also infected. As with cankers, the spores spread fastest and farthest during wet weather, so it is best to avoid shearing on damp days. Sterilize knives or shears in a chlorine bleach solution after working on infected trees. Certain other diseases, such as the rusts, also cause needle drop, so are often confused with the needlecasts.

Dothistroma needlecast strikes Austrian pine and is identified by the browning of the needle tips just beyond a reddish brown band. Fungicides applied during the growing season and once after growth stops can help control it.

Lophodermium needlecast is hosted by Scotch and red pine. It discolors harvest-sized trees enough to make them unsalable and often kills seedlings outright. Symptoms to look for include brown spots with yellow margins on the needles in spring and an abundance of brown needles at the base of the tree later in the season. To help control this needlecast, apply a fungicide to infected trees every three weeks starting in late July, continuing until October.

Naemacyclus, or *Cyclaneusma, needlecast* is most often found on Scotch pine. The symptoms are light green spots on older needles in early fall, followed by yellowing, and finally browning of both old and new needles. Three sprays of fungicide at two- to three-week intervals beginning in mid-April help control it, but 100-percent control is seldom possible.

Rhabocline needlecast affects Douglas fir, and the Rocky Mountain strain is especially susceptible. Yellow spots show on infected needles in late fall, and by early spring the needles begin to turn yellowish to reddish brown. The disease can be spread by pruning tools, so shear all your healthy trees first, and disinfect your tools in full-strength chlorine bleach after pruning infected areas. This needlecast is difficult to control, but it helps greatly to remove any badly infected trees and to spray fungicides every week from the time the buds begin to swell in the spring until they fully open.

Rhizosphaera needlecast chooses spruce trees, especially the Colorado spruce, for its dirty work. Part of the needles turn purple-brown and drop in late fall. Spray when the needles are half elongated and again after they are fully elongated.

Swiss needlecast attacks Douglas

fir. In midsummer, the needles turn brown, then fall off by early fall. Use a hand lens to check for rows of black, fuzzy, fruiting bodies and tiny openings on the underside of the needles in early spring or late fall. Spray fungicide twice in early spring, about three weeks apart.

Root Rot. *Armillaria,* or *shoestring root rot* as it is sometimes called, is a fungus that attacks the root collar (the part of the trunk just above the roots). It can strike any species of evergreen and kills by girdling. Watch for yellowing followed by browning of needles, resin on the bark at the root collar, and creamy white fungus under the bark at the root collar. Control is difficult. Keep the trees in healthy condition so stress won't invite the disease, and remove any infected trees, including the stumps.

Rusts. Various kinds of rusts strike evergreens, and they have the unique habit of spending alternate times of each year on a completely different species of plant. Apple growers are familiar with the *cedar apple rust* that alternates between red cedar (juniper) and apples; and currant and gooseberry growers are quite aware of the *blister rust* that their plants can spread to neighboring white pine trees. Those who raise balsam fir have to watch out for *uredinopsis rust,* a needle-dropping disease that alternates between balsam and the sensitive fern *(Onoclea sensibilis). Pine needle rust* strikes red and Scotch pines, alternating with both goldenrod and the native fall-blooming asters.

The symptoms of rusts on red and Scotch Pine are orange blisters on the needles; on red cedar, you will find brown, warty galls up to 2 inches in diameter on the twigs. On white pine, watch for patches of brown bark with yellow borders, spindle-shaped swellings on the branches and trunk, open cankers with yellow borders, and reddish brown needles on dead branches and tree tops. Blister rust can spread fairly rapidly throughout a tree and finally kill it. On balsam fir, the uredinopsis shows up first as white spores on the underside of the new needles in early summer. The spores then scatter to nearby sensitive ferns, and the needles turn yellow and finally drop off in late summer. The disease overwinters on the ferns, then spreads again to the firs during wet weather in early spring.

Best control of the rusts is to remove the alternate hosts. Clean out all wild gooseberry and black currant bushes within 900 feet of white pine plantings. If you grow balsam fir, mow all sensitive ferns, or spray them with an herbicide before mid-August when the spores return to the trees for the winter. Eradicate neighboring goldenrod and aster if rust diseases shows up on your Scotch and red pine.

Insects That May Bother Christmas Trees

The long list of insects that can infect Christmas trees could be frightening, but because evergreens are remarkably insect-resistant, your trees may never be seriously bothered by any of them. Some of these pests are always present in a forest, but seldom in large numbers. Others inhabit only small sections of the country, and a few specialize — they live in only one species of tree. The following list is intended as a reference to help you identify an insect attack. Examine your plantation frequently for invasions, because when conditions are right, bugs have a way of multiplying faster than even Noah intended.

The two best times to spray for insects are when the adults first become active and are laying their eggs, and when the young larvae have just hatched and begun feeding. *Bacillus thuringiensis* is one of the safest products that can be used to destroy chewing larvae, but a contact poison, such as malathion, may be necessary to control aphids, mites, and other sucking insects.

Ants. Mound ants kill off trees in some parts of the country. They also build huge hills in plantations, which interfere with mowing and the movement of equipment. Ants also culture aphids much like dairy farmers raise cows; but instead of milk, ants produce from them a sweet, sticky, honeylike substance as they "pasture" the aphids from tree to tree. Control ants by soaking their hills with an insecticide mixed with water, preferably just before a rain. Use a generous amount so you are sure the queens will be killed. In our Christmas tree lot, we sometimes find that bears have ripped open the hills to eat the ants, but that is not a recommended way of eradicating them.

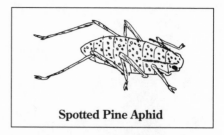

Spotted Pine Aphid

Aphids. These little plant lice come in various shades of green and black or sometimes spotted. They suck the juices from needles and bark, weakening a tree without leaving any obvious signs of damage. The *balsam twig aphid* attacks balsam fir trees in summer, causing the needles to become twisted, exposing the blue-white underside, and badly deforming the tree. *Cinera aphids* choose Douglas fir for their favored diet, and *spruce gall aphids* cause pineapple-shaped galls that disfigure the twigs of spruce branches. Various kinds of pine aphids attack both bark and needles of pine trees. Although aphids are small, they can be seen during close inspection. Small amounts of aphid damage can be sheared away, but if damage is widespread, this is impossible. Spot spray with a contact poison as soon as the damage is noticed. Overfertilization and other forms of stress that weaken a tree are leading causes of excess aphid activity.

Bagworms. If your trees contain silky, weblike bags with needles on them, you can be sure that they were built by bagworms for raising their next generation. They attack spruce, fir, and white pine, and can devastate a tree in a short time. Picking off the bags and burning them before the

Bagworm

eggs hatch helps to prevent their spread, and spraying weekly with *Bacillus thuringiensis* from late May through June helps control those that hatch. It helps to get rid of any nearby red cedar that serve as hosts before planting your trees, too.

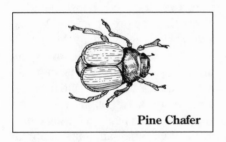

Pine Chafer

Beetles. Although a huge number of beetles stalk the earth, most are harmless to conifers, and some, like the ladybug, are beneficial because they eat aphids. Not so helpful though, are the *pine chafers,* which chew the needles of all pines, and *pine bark beetles,* which bore into the bark of all species of conifers. Spraying controls the chafers, but bark beetles and borers are difficult to get at, so it is best to remove and burn all infected trees.

Budworms. Budworms usually are more deadly to large evergreens than to those of Christmas tree size, but in one of our young plantings, the larvae blew in from a nearby forest and completely defoliated the trees before we were aware of it. The *jack budworm* attacks Austrian, red, and Scotch pines, and the *spruce budworm* has devastated many acres of fir and spruce in Eastern Canada and northeastern United States in recent years. The larvae spend the winter underground, come out as moths in the spring, and lay their eggs soon after. Spraying with *Bacillus thuringiensis* in early summer, just as the larvae begin to feed, gives good control.

Grubs. The larvae of the *May beetle* sometimes feed on the roots of seedlings, killing or weakening them. If they are present in your planting, treat seed and transplant beds with a granular insecticide each spring. The grubs prefer grass roots, but will eat newly set out trees if no grass is present. If your field is mostly in sod, the trees are not likely to be attacked — good reason not to eliminate your grass cover completely.

Leafminers. Leafminers, or *needleminers,* tunnel into the interior of needles, then form the dead needles into webbed clusters. They attack all species of spruce and can disfigure trees so badly that they become totally unsalable. Spray in early summer just as the larvae emerge from the eggs.

Midges. The tiny *balsam gall midge* forms abnormal growths (called galls) on the new sprouts on balsam and Fraser fir needles, causing them to drop off. Spray in late spring with a systemic or contact insecticide for the best control.

Mites. *Spruce spider mites* attack all species of Christmas trees, discoloring and sometimes killing them. They cause yellowish to rusty-brown shoots, with fine webbing between the needles. *Eriophyid mites* are tiny, cream-colored creatures that attack pine trees, causing yellowing and twisting of the needles. To control, spray with a miticide beginning in early summer, repeating every ten days until new damage stops appearing.

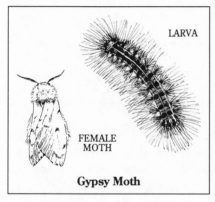

Gypsy Moth

Moths. A large number of moths frequent conifers. The *Adana tip, European pine shoot* and *Nantucket pine tip moths* attack the buds and new growth of the various species of pine trees and deform them. *Pine tube moths* choose white pine; the *tussock* feeds mostly on white and red pines, but sometimes spruce and fir; and the Nantucket moth goes for Austrian, red, Scotch, and Virginia pines. *Zimmerman pine moths* attack all pines, but prefer the Austrian and Scotch. *Gypsy moths* ordinarily strip the leaves of deciduous trees, but when their numbers are large, they also assault evergreens.

To control moths, spray an insecticide as soon as you see the larvae start to feed.

Nematodes. These tiny creatures feed on the roots of plants and are able to reproduce with unbelievable speed. The *pine wood nematode* feeds on Scotch pine and occasionally other young pines, often killing them. Nematodes are found in all parts of the country, but are more common in the warmer areas. Symptoms are a yellowing of the needles, which eventually turn brown but remain on the tree.

If you know that nematodes exist in your area, sterilize all seed and transplant beds before planting, and avoid planting trees in dry locations. Nematodes are almost impossible to control in existing plantations, and in badly infected areas, the only solution is to cut down the trees, sterilize the soil, and start over.

Sawflies. Sawflies are one of the more obvious insects found on Christmas trees. Even with their color camouflage, they are large enough to spot easily, and their damage shows as soon as they start feeding. The *European sawfly* attacks Austrian, red, Scotch, and Virginia pines, eating the older needles and giving the trees a thinned out look. The *redheaded pine sawfly* feeds on all ages of pine needles, often stripping the whole tree. Usually the damage occurs in cycles, and after one bad year, no more may appear for a decade or more. For

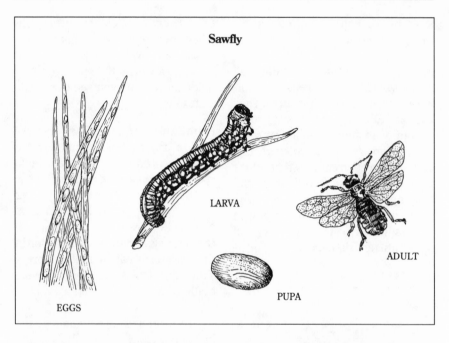

Sawfly

LARVA

ADULT

PUPA

EGGS

good control, spray with an insecticide as soon as the feeding begins and continue at intervals until it ends.

Scale. Many people mistake scale for a disease, because the insects are so tiny you need a glass to examine them and their damage appears scalelike on the needles or bark. These insects not only disfigure the trees, but they also weaken it by sucking out the sap. *Pine needle scale* is found on all species of pines and spruces and on Douglas fir. Look for white-flecked or brownish needles, often with light-colored, oyster-shell-like growths on them. *Pine tortoise scale* strikes Austrian, red, and Scotch pines in much the same fashion, except that the needles turn a sooty, shiny color, and the scales are reddish-brown and helmet-shaped. *Spruce bud scale* attacks all spruce

species, discoloring trees and sometimes killing them. Symptoms are dying shoots, and red-brown, globe-shaped scales at the base of the new growth. Any good insecticide applied as soon as the damage begins helps to control these pests. Remove and burn any badly infected trees.

Spittlebugs. Everyone is familiar with the spittlebug, which leaves a saliva-like material on grass and other plants in the summer. A similar creature also invades trees. The *pine spittlebugs* attack most evergreens, but unless they appear in large numbers, seldom do much damage. The *Saratoga spittlebug* feeds on the needles of pines and occasionally firs, discoloring and sometimes killing them. Besides the obvious spit, look for red-brown needles and tan flecks under the bark of older wood. The alternate hosts are ferns, brambles, and broad-leaf weeds, so mowing or treating these with herbicides helps greatly to control the insects. Spraying an insecticide in midsummer also gives good results.

Thrips. Although these insects are not usually a major concern to tree growers, sometimes they build up a large enough population to devastate Austrian and Scotch seedlings and young trees. Look for crooked, curled, discolored needles, and needles of varying sizes with brownish wounds on them. Because thrip damage is likely to be worse during dry periods, it helps to irrigate seed and transplant beds and to spray an insecticide in late spring before the eggs are laid, as well as two or three additional times about a week apart.

Webworms. All pines and red cedar trees can be bothered by webworms. Check for masses of brownish needles, insect waste held together with webbing, and needles that have been eaten. Larvae are from a half-inch to nearly an inch in length. The best control is to remove and burn nests before the larvae leave them, or spray the larvae if they are already feeding.

Weevils. Several types of weevils (beetles with long snouts) bother conifers. Because many species breed in weak and dead trees and in old stumps, remove and burn all those that are near your evergreens for best control. The *northern pine weevil* feeds on all pines and spruces. Look for discolored and twisted shoot tips

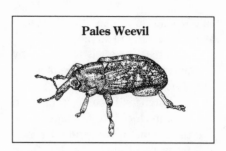

Pales Weevil

and for light brown, white-spotted weevils less than a half-inch long feeding on the shoots after dark.

Pales weevils feed mostly on pines and Douglas fir, but may be found on spruces also. They are especially devastating in older plantations where harvesting has left many stumps. Watch for dead shoots and wounds with pitch oozing from them, and, after dark, look for dark-colored weevils about a third of an inch long.

To control both Pales and northern pine weevils, drench infected stumps with an insecticide in early spring. Spray living trees in late summer while the weevils are feeding.

Pine root collar weevils feed on pines, particularly the longer-needled species, girdling the trunk at the point where the roots grow from it, and usually killing the tree. Symptoms are yellow to red needles all over the tree, black pitch around the base of the trunk, and small, yellow-white larvae tunneling in the bark and surrounding soil.

To prevent weevil damage on Scotch pines, plant the more resistant strains such as the French or Turkish. Drench the soil around each tree in infected plantings with an insecticide during warm weather in late spring, to kill adults. Repeat in mid-August to kill larvae.

Pine root tip weevils, found mostly in the Great Lakes states, attack the root tips of pines, sometimes killing them. Watch for discolored and dead needles, and for white, C-shaped, half-inch larvae on the root ends. Drench the trees and the soil around them with an insecticide recommended for weevil control. When harvesting an infected plantation, cut the trees close to the ground.

White pine weevils are found mostly on white pine, and Norway and Colorado spruce, but other pines and firs may be affected to a lesser degree. Damage is likely to show first at the top of the tree, where all the needles may be dead. The top leader will curl downward, and pitch flowing from it will be obvious. White

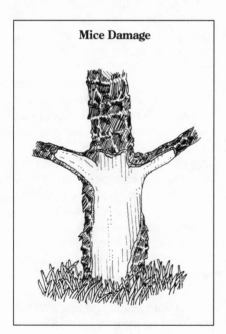

Mice Damage

cocoons appear in early to midsummer, followed by brown weevils that feed on the tree. Prune back dead tops whenever they appear. Soak the tree tops with an insecticide as soon as the weather warms in the spring and again in summer when the weevils appear.

Animal Damage

Animals from mice to moose can be tough on evergreens. Both wild and domestic creatures take their toll, and identifying the culprit isn't always easy. Deer and elk damage can usually be readily spotted, as they chew the bark of young trees or rub their antlers on larger ones. Other gnawings on the bark, such as that of mice and voles, can take place under the winter snow and come as an unpleasant shock in the spring, because if a tree is completely girdled, everything above the girdle will die. Squirrels, grouse, and pine grosbeaks often eat the terminal buds of evergreens during the winter months, causing irregular growth in the spring. Even beavers and porcupines sometimes eat the bark of conifers, and occasionally cut off limbs, although their preferred foods are the hardwoods. Birds, including cute little songbirds, often land on the fragile tops of evergreens in early summer and break them over.

Domestic animals such as cattle, horses, goats, and sheep have the habit of breaking out of their confinements from time to time and heading for an area where they can do the most damage.

Even if you successfully identify

the guilty party, control, especially of the woodland creatures, can be difficult. Some growers report having successfully used animal repellent sprays on their trees, but we've had no such luck. We found that even though a deer might not have liked the taste of one branch, she bit off other samples just to be sure, and all her relatives behind her were doing the same thing. Using poisoned oats or corn will cut down on mice damage, but if you use them, place them under boards or in cans so birds won't get them, or else spread them late in the fall so they will be covered with snow.

You may want to encourage the hunting of porcupines and squirrels if they are hurting the trees, but check your state laws or extension service before shooting game animals. In some states, Christmas trees are not considered a crop under the fish and game laws, so rabbits, deer, and elk cannot be shot for doing damage in a tree plantation (although it might be legal if they were destroying an orchard). In any case, it may be good to discourage the taking of wolves, fishers, foxes, coyotes, bobcats, and weasels in the vicinity of your trees. They, plus hawks, owls, and eagles, may be your best friends when it comes to controlling animal pests.

Human animals sometimes damage trees, too. Cross-country skiers and persons riding snowmobiles, all-terrain vehicles, and trail bikes often inadvertently run over young trees, especially if they are hidden in the snow. Signs and fences help to discourage these activities until the trees are big enough to be noticed.

People can also be problems after the trees reach harvest size. Treenappers range from naive, otherwise solid citizens who innocently think all trees are a gift of nature and free for the taking, to professionals who load tractor trailers on moonlit nights. One December I found a shiny new Oldsmobile parked by one of our lots and spotted a woman in a mink coat busily sawing down a tree. After congratulating her on her good taste, I asked if she planned to steal her Christmas turkey, too. Furious, she screamed that she didn't know the trees (even though they were carefully spaced and sheared) belonged to anybody. I wondered how she would feel if she found me cutting down a blue spruce on her suburban lawn.

Theft is difficult to control. Signs may alert people that the trees are not to be cut, but thieves have been known to carry off the signs as well as the trees. Growers sometimes border their plantings with thorny hedgerows of rugosa or multiflora roses (these can turn into vicious weeds as birds spread the seeds) or install strong fences to make robbery more difficult. Continuous patrolling may help, but it is time-consuming and difficult. Many plantings are far from a dwelling and anyone who has worked hard all day hardly feels like patrolling all night in freezing weather. One man I know sleeps in a cabin overlooking his plantation during the weeks before Christmas. His German shepherd stands guard and a high-powered rifle lies nearby, ready to fire a warning shot to discourage any illegal pre-Christmas shopping.

I wouldn't advise accosting a tree-napper at gunpoint, however — instead of stopping a crime, you may find yourself accused of assault with a deadly weapon. Instead, try to get a license number or other identification and notify the police so they can nab the miscreants with the evidence. Painting the tree stems with an identifying tree paint has also been tried with some success, but as with putting identifying tacks into some of the trees, the thieves have to be caught with the goods before the evidence can be used, and that is often difficult.

CHAPTER FOURTEEN

Managing for
Efficient Production

Christmas trees are quite unlike most other crops. If you raise vegetables, grain, or cut flowers, you can harvest them the same season. With fruit or nut trees, it will be nearly a decade before your first big crop, but at least that harvest is not your only one. After you plant Christmas tree seedlings, however, you have to wait eight to twelve years, and after a harvest, you must plant again and wait for the next crop.

Since the time from planting to harvest is so long, growers have developed various ways to shorten the growing period and produce annual harvests of a predictable number of trees. By raising faster-growing species, using better weed control, and fertilizing more scientifically, they can shave several years off the length of time necessary to grow a crop. Furthermore, most serious growers no longer wait until all the trees are harvested before replanting. Instead, they set a four-year-old transplant between each of the larger

trees a few years before they begin cutting. Some even find it worthwhile to produce several crops of trees from the same set of roots.

Interplanting
Planting a new transplant between each of the larger trees two or three years before they are cut is a common practice in many plantations. This interplanting means the land will not be wasted during the several years when only one tree would otherwise be growing, and shortens the time before the next harvest. The two or three years will allow the young tree to grow big enough to be seen and therefore not be damaged when the mature trees beside it are cut.

Interplanting is not always practical. Sometimes it is best to wait and replant only after all the trees are cut, for example, where diseased or insect-infested tree stumps or large amounts of brush are present or when replanting must be done with a machine. In such cases, clearing everything out and starting over will be the best practice.

Often, however, setting a new transplant between trees that will soon be sold may be worth the effort, even though the planting must be done by hand. The same species of tree may be used for replacements, or there may be a good reason to switch to strains or species that are more pest resistant, more marketable, or faster growing.

A good-sized transplant (2-2 size at least) is best for interplanting, because it will be crowded in with the larger trees and thus needs to be rugged enough to fend for itself. To give the transplant a better chance of thriving, get rid of weed and grass competition. One way of doing this effectively is to spray Roundup on the spot the fall before planting the following spring.

When interplanting, always provide adequate fertilizer to the new trees to compensate for the nitrogen that is already being depleted from the soil — first by the competing larger trees, and after harvest, by their decomposing roots. Additional weed control is usually also necessary, because weeds will grow faster thanks to the extra fertilizer and additional sunlight they'll get after the larger trees have been cut.

Stump Culture

Most young evergreens have the ability to regenerate from their roots even after they have been cut back severely. Some growers take advantage of this characteristic and raise a new crop of trees on the same stumps

in about a third of the time it would otherwise take, without the expense of buying and planting seedlings. When all goes well, a completely new tree can be produced in as little as five years after the original one has been harvested. The process is called stump culture. It works because the harvesting of Christmas trees comes when the trees are completely dormant and can therefore stand this severe shock. If the same tree were cut nearly to the ground in summer, it would probably bleed badly and die. Stump culture is most common when growing firs and spruces, but it is seldom done on large tree farms because it does not fit in well with mechanical culture. It can be useful in small plantings or in areas where land is scarce and expensive, but is practical only when a planting is free from diseases and insects that live over in stumps and roots and are likely to infect subsequent crops.

Planning for this type of culture must begin early in the tree's life. When you are doing the first basal pruning, leave two or more of the bottom branches just below the bare stem you have left to serve as the "handle" of the cut tree. When the tree is harvested, make the cut just above these branches and leave them growing on the stump. In a year or two these branches will start to turn upward.

The fall after this upward growth occurs, cut away all the growth except the most vigorous, upright-growing branch. With your help, this will become the new tree, although it won't look much like a tree at this stage.

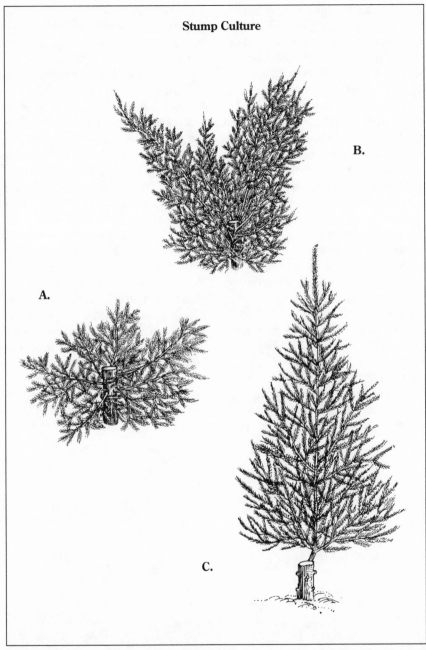

Stump Culture

B.

A.

C.

When harvesting trees, leave two or more branches on the stump below the cut (A). After one or two years, (B) these branches will begin to turn upward as they grow; the next fall (C) cut away all but the most vigorous branch.

When it starts to grow the following spring, shear it lightly so the branches will start to grow evenly all around. Continue to shape it each year by shearing it during the growing season just as you would any tree.

The tree will grow rapidly because there is so little top in relation to the roots, so the first few years be stingy with plant food. Stimulating even faster growth will result in a weak tree that could break off easily where it joins the stump. Usually staking is not necessary, but if the new tree does not grow straight or appears weak, tie it to a stake for the first year or so.

By the fourth year, when the top growth has begun to catch up with the roots, a half cup of 10-10-10 fertilizer on each side of the tree each spring will help it sustain good growth and improve its color. If you want to try for a third crop using stump culture, basal prune the new tree the same as you did the original, leaving two or three bottom limbs in place.

Generally, three or four trees can be grown in this way before the stump becomes too large to ensure good results.

Even though stump culture is not always practical, it is interesting to work with this regeneration process. Like trying for the first ripe tomato in the neighborhood, it can be a challenge to see how short a time it takes to produce a good-sized tree.

So that you don't waste a lot of time experimenting, before you choose a method of replenishing your plantation, visit other growers to find what is being done in your locality with the species you are growing. With greater competition and increases in land prices and labor costs, it becomes imperative to produce the best trees in the most efficient way, whether by interplanting or stump culture. It isn't difficult to figure out the advantages of growing six or eight crops during your career rather than only four or five.

CHAPTER FIFTEEN

Making Money with Greens and Wreaths

We used to call it brush, but the foresters taught us to say greens. You can't sell brush at 15 cents a pound; it sounds too much like a waste product. Greens and greenery, on the other hand, suggest fragrant Christmas wreaths, sprays, garlands, and centerpieces. Even during years when trees have been in oversupply in certain areas, there have seldom been too many greens and the products made from them. Wreaths and roping are hung almost everywhere. They brighten large mansions, tiny cottages, workplaces, shopping plazas, and city lampposts, and we even see them hanging on such unlikely places as barns, silos, automobiles, buses, and cemetery stones. Unlike trees, some people leave their wreaths on display until Easter, and since they are usually hung outdoors, they are not affected by the fire regulations that often forbid indoor trees.

Greens are an excellent sideline to the Christmas tree business; in fact, some growers find that they make more profit from their greens and wreaths than from their trees. Very little investment in equipment and supplies is necessary, and you don't even need land to get into the wreath and roping business. Some people we know produce hundreds of wreaths each year and don't raise a single tree, but instead buy all their greens. Others gather branches left over from timber-cutting operations or nearby Christmas tree harvests.

Firs and pines are the species most often used for this purpose, either separately or in combination. The spruces are too prickly, which makes them difficult to fashion into wreaths and roping, although they are often used for making the heavy roping used for street decorations and the grave blankets used in many cemeteries. Cedars, yews, and junipers are less useful because most tend to lose their rich green color in early winter.

Where they grow well, other popular materials for holiday greens are the broadleaf evergreens such as bay, boxwood, holly, and laurel. Native club mosses, including ground cedar, princess pine, and reindeer moss are sometimes used, and many of these are better suited for indoor use since they don't fall apart in a warm room. Some may be in short supply or even classified as endangered species in certain areas and consequently it may be illegal to gather them from the wild. Even if they are not endangered, always cut, rather than pull them, so the roots will remain to produce more greens.

Producing Greens

In this seasonal business you are entirely dependent upon the weather for your start-up time in the fall. The trees need near freezing temperatures to "harden" the needles so they won't fall off a few days after the branches are cut, and thus, cutting cannot begin until it has become cool. Greens used to be cut entirely from forest trees, often along with a pulp-cutting or lumbering operation. Usually only the greens in the middle portion of the large trees were desirable. The ones at the bottom grew in such heavy shade that they were thin, and those at the top were exposed to so much wind that they were coarse.

To harvest greens from cut trees for the wholesale market, cut the boughs in lengths of about 30 inches and tie them in bundles of approximately 50 pounds each. These are sold by the ton or half ton. Branch diameter should not be much over a half inch at the cut end, and each part of the branch should be completely

covered with needles. Some people use hand pruners for cutting, but most hold the branch over a log and cut it to the right length with an axe.

Tip greens — the new growth at the ends of branches — are often harvested either for the wholesale or retail market, or for one's own processing. They are sometimes cut from standing trees with clippers or pole pruners. They are usually a foot or less in length and are wholesaled either in bags, packed between longer lengths of evergreens and tied in 50 pound bundles, or baled in a baling box. These tips have very little waste in them and are sold at much higher prices, often by the pound.

After you have cut the greens, they must be kept cool and moist or the needles will dry and fall off. A cool, rainy fall may not be comfortable for you when you're cutting, but it is ideal for producing high-quality greens. Windy days, as well as warm, sunny weather, can dry out and ruin cut greens quickly, so store them in a cool, sheltered, shady spot, and sprinkle them occasionally if necessary. Don't pile them over 3 feet high or the bottom ones will be crushed. They may be covered with burlap, branches from unusable trees, or a similar material, but don't use plastic because it can cause the greens to get hot, ferment, and spoil.

Planting and Managing a Greens Orchard

If you make wreaths and other holiday decorations, it is often difficult to find reliable sources of good greens. To be sure of a supply each year, many wreath-makers grow their own pine, fir, boxwood, or holly in an "orchard" that is planted and cultured entirely for the production of high-quality greens.

My original greens orchard was created by cutting off the tops of wild trees and thinning them so that they were about 8 feet apart. There was nothing wrong with the greens from these natural trees, but I eventually switched to gathering them from planted trees, because the wild ones were so far away that it was difficult to harvest them easily.

We have about 3 acres of balsam fir in our greens plantation. When they were originally planted I intended to grow them for Christmas trees, but the deer chewed them so badly that they became disfigured and couldn't be sold as trees. My rage at the deer diminished somewhat after I realized that growing greens was even more profitable, because I didn't have to sacrifice the tree every time I harvested.

The trees had originally been planted 5 feet apart, which is too close together for greens trees, so they eventually had to be thinned enough so they wouldn't crowd each other. An original spacing of about 8 feet each way would have been best. We grow our balsams in an exposed location in full sunlight, which makes the twigs extra bushy. Some people prefer to have less dense greens with a softer texture and they plant their orchard on the east or northeast side of a row of tall trees so they are protected from wind and get only a half day of full sun. Pines grown for greens

should get sun all day, however.

When the trees reached about 4 feet in height, I cut off the tops to force more growth into the lower branches and the next year began clipping them lightly. By dividing the plantation into two equal sections, I alternate the harvest by clipping greens from one lot one year, and the other the next.

Clipping stimulates the tree to develop so many extra buds that the new growth from each one is shorter than normal. These thicker, shorter twigs are ideal for making wreaths and roping. In late winter every year I cut off the top of each tree to keep them from growing out of reach. Even with this severe treatment, they still grow a little, so every ten years or so we have to cut the tree back to about 2 feet in height. This rejuvenates the tree and prevents the needle growth from becoming hard and coarse. By cutting back only part of the trees each year, the annual production of greens is never greatly diminished.

The process of gathering greens is a lot like shearing sheep, except that it is done in late fall instead of spring and only biennially. I cut off about 10 inches from the tips of the branches with hand pruners. This length is best for wreaths made on 16-inch rings or larger, but we must cut them into shorter lengths as we make smaller wreaths and garlands. If we were selling them as greens, we would bundle or bale them in the field, but since we use them ourselves, we pack them into large grain bags, which makes them easy to handle. A full bag weighs about 40 pounds, and from it we can make approximately seven 12-inch-ring-sized wreaths. I like the bags made from woven plastic better than those from burlap because sitting on the wet ground doesn't rot them as it does the burlap. After filling the bags, I leave them in the field overnight. Then because the greens keep better when they are cool, I move them to a shady spot out of the wind in the morning when they are damp and frosty.

I fertilize the greens trees very little because they were planted on a field of good soil and cutting off the tips once every other year doesn't greatly deplete their energy. I check them during the summer, though, and if they are not a rich green I scatter a pound or two of 10-10-10 around each tree. This quickly improves their color. I do not mow the greens orchard and used herbicides only the first two years after planting, to keep competitive growth suppressed. The dense-growing trees compete satisfactorily with the heavy grass cover, but I must occasionally cut out wild apple or poplar seedlings.

The greens orchard has worked out beautifully and I recommend it for anyone who has a place to have one. It's a relief not to be searching for new sources of greens each year, and it is a labor saver to have them close to our home.

The Wreath Business

Many entrepreneurs make only a few wreaths to sell locally, but in evergreen country, making and selling plain and decorated wreaths is often big business. Many are wired by hand at home by families who may make several thousand wreaths and a good portion of their yearly income in a few weeks of intensive work. The younger children in the family cut the greens into the proper lengths and place them in bunches of three or more and the older folks do the wiring. Some large, commercial wreath-making enterprises contract for people to make wreaths in their own homes, often furnishing the greens and wire; others hire a number of wreath-makers to work together in a large shed in factory-like assembly lines.

Not all wreaths are made by hand. There are currently two types of wreath-making machines on the market. One uses a special type of ring with clamps surrounding the perimeter. The ring is placed on a special table and a handful of greens are laid on the ring. Then, when the operator presses a foot lever, one clamp is bent over, tightly fastening the greens in place. This process is continued around the ring until the wreath is complete.

Because this machine is simple and no electric power is needed, it is inexpensive. Rings of a special type are required, however, which are more costly than the simple corrugated ones used for making hand-wired wreaths. Only a single-faced wreath can be made on these machines, too, so their use is limited, since many buyers prefer those that are double faced.

The other type of machine is more complicated and expensive. It is powered by an electric motor, uses winding wire and ordinary rings, and the process duplicates more closely the making of a hand-wired wreath. A bunch of greens is held against the ring, then a foot switch turns on the power, and a spool of wire rotates around the ring, wiring the greens to it. By alternating the placement of the bunches, the operator can make a double-faced wreath. The machine can also be used to make garlands by wiring small bunches of greens onto a heavy string.

With practice, it is possible to make an attractive wreath quickly with either machine, but many buyers still prefer those wired by hand. Skilled hand wirers can outdo either type of machine in both speed and quality. Some can produce six or eight wreaths of the 12-inch-ring size each hour in an eight-hour day.

Although most wreaths are made by wiring greens onto a ring, florists sometimes create them by sticking sharpened evergreen twigs into rigid foam forms made in circular and other shapes. Instead of using commercial forms you can make your own by packing sphagnum moss around a double hoop, moistening it and wrapping it with burlap. These keep the greens fresh when they are used for centerpieces, mantelpieces, and similar decorations.

Setting Up the Workplace

It is important to work in a cool room, even if you have to wear a heavy sweater. Greens dry out rapidly and the needles fall off if the temperature is warm. Try to set up your workplace where there are adequate windows to allow in lots of natural light, because brush color is hard to see accurately by artificial light (although the daylight-type fluorescent bulbs are very good).

Wherever you work, be comfortable. Make sure your table is set at absolutely the right height so it won't be necessary to bend over, even a little. Plan to change positions occasionally to avoid getting stiff muscles. If you sit as you work, choose a comfortable chair or stool; if you stand, use a mat to stand on. Organize your workplace so the wire, rings, greens, clippers, and string to tie the wreaths into bundles are all close at hand. A radio or a tape deck can provide relaxing music, let you listen to books on tape, or study a new language. I like to take a break after each ten or fifteen wreaths and reward myself with a cup of coffee, an apple, or other treat.

Both wire and evergreen needles can be hard on hands, so many people like to wear tight-fitting cotton gloves when they work. I've found that one pair is good for about 200 wreaths!

To make the most money in the wreath business, speed and efficiency are very important. There are only a few weeks in the fall between the period it is safe to start cutting the greens and the time the wreaths are needed. If you are new at wreathmaking and plan to do it as a business, it is worthwhile to practice making a few dozen before the season actually begins. Try to develop quality first, then work on improving your speed. Those first creations will dry out before the holidays and must be discarded, but write it off as tuition. Don't be discouraged. If your first attempts appear to be square or egg-shaped, if they fly apart when you shake them, or if each takes an hour to make, it's par for the wreath-making course.

Of course, if you make only a few wreaths, time is not too important, but if a great many are produced, each minute saved makes a big difference. If you are making a thousand wreaths and spend five minutes more than necessary on each, it will add up to a whopping eighty-three hours, or more than ten 8-hour days!

To make your time and labor most profitable, find out the type of wreath your buyers want before you begin. The wreaths made in our area vary greatly in weight, quality, and price. One woman makes only a few extremely beautiful, bushy ones that she sells at high prices to what used to be known as the carriage trade. Another family makes a huge number of light, lower-quality wreaths that go into special markets at bargain prices. Both families make good money and are satisfied, but both are understandably somewhat contemptuous of each other's business.

Supplies Needed
Order your supplies long before the wreath-making season begins. Mail-order houses always have a mad rush in late fall and may not be able to furnish everything you want at the last minute. If you buy from a local supplier, try to buy early, too, because most retailers don't want to hold over seasonal materials until the

following year, and seldom order more than they think they can sell.

Good-quality hand pruners are most important. If you cut your own greens you probably already have these on hand, but they may need sharpening. If they are not sharp, your blistered hands will soon let you know. It's easy to break or lose clippers, so I always keep an extra pair on hand.

Crimped wreath rings come in a variety of sizes from 8 inches to 24 inches in diameter, and smooth heavy rings are available in even larger sizes. The 12-inch ring is the most common, and a good, finished, double-faced wreath made on it should measure about 18 inches in diameter and weigh from 4 to 5 pounds. The 8-inch ring should result in a wreath 14 inches in diameter; the 16-inch ring is used for one of 24 inches, a 20-inch ring for a wreath of about 28 inches, and one on a 24-inch ring should have a diameter of about 34 inches.

Winding wire is usually purchased on half-pound spools, in 22-, 23-, and 24-gauge sizes, and in either natural wire color or green. We like to use the green because it is less visible on the wreath or roping. Most people use the lightweight, 24-gauge size, because more wreaths can be made per spool, but if you make large wreaths you may break the lighter wire now and then and waste so much time splicing it that the heavier 23- or perhaps even the still heavier 22-gauge will save you a great deal of time. A 23-gauge spool makes approximately eight to ten wreaths of the 12-inch ring size.

How to Make a Wreath by Hand

Single-faced wreaths, made by wiring the greens on only one side of the ring, are suitable for hanging on a wall or door, or to use as a centerpiece. Double-faced wreaths, with greens on both sides of the ring, are preferred by most people because they are more bushy and, since the wreath wire is hidden, they can be hung in windows.

Since the upper and lower sides of greens such as fir, holly, and boxwood look quite different, they must be "faced" when placing them on the ring so the pale side won't show. Pines look the same on both sides, so this facing is not necessary.

To make the wreaths, follow these steps:

1. Wind a few twists of wire around the ring to fasten it securely.

2. Place a bunch of two to four tips of greens on one side of the ring and wire the base of the bunch to the ring with two or three tight wrap arounds. Select some good bushy greens for this first bunch, because it must hide the base of the last bunch you'll insert. If you are making a double-faced wreath, flip the ring over, and wrap a similar bunch onto the back side. Place it almost, but not quite, opposite the first one.

3. Lay another bunch of greens over the base of the first, hiding the wire, and wire this one to the ring. Continue in this fashion, moving around the ring and placing all the greens on one side if it is a single-faced wreath or flipping it for each bunch if it is a double.

4. When you reach the spot where you began, tuck the base of the last bunch underneath the tops of the first that you wired. Wire it in carefully, so neither the stems nor the wire show.

5. Cut or break the wire and fasten it tightly with several twists to one of the wires or to the ring itself.

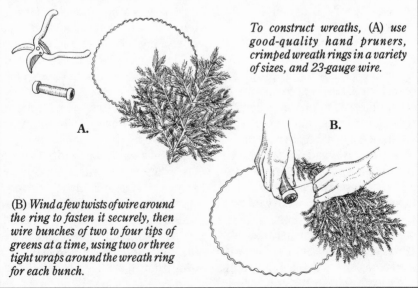

To construct wreaths, (A) use good-quality hand pruners, crimped wreath rings in a variety of sizes, and 23-gauge wire.

A.

B.

(B) *Wind a few twists of wire around the ring to fasten it securely, then wire bunches of two to four tips of greens at a time, using two or three tight wraps around the wreath ring for each bunch.*

If you make a lot of large wreaths, and have the time, you may want to buy heavy wire from a salvage company and make your own rings. Another economizing tactic is to buy winding wire of the right gauge on large coils and rewind it on used spools with an electric drill.

You'll need to order heavy string or baler twine to tie up the bundles unless your buyer wants them loose so he can inspect each one. Wreaths were once sold by the dozen, and six were tied in each bundle. Five in a bundle is more common now, and it's much easier to count them. I tie the 20-inch wreaths in bundles of three, so they are lighter to load onto a high truck. Any wreaths 24 inches or larger, we don't tie at all, for the same reason.

If you decorate wreaths, you will also need ribbons, cones, wired picks, and berries (either real or ersatz). As you scan a supply catalog, you'll probably find many more things you can use (see Appendix for addresses).

Storing Greens and Wreaths

Evergreen boughs, roping, and wreaths must be kept fresh until delivery time, so I always hope for a fall season that is cold and damp. Keep them piled out of the sun and wind, or they will brown and dry out quickly. We store ours in a shaded, sheltered corner where two buildings come together, but even then we sometimes have to sprinkle them with water during dry periods. Water them only in the evening or early morning, when the greens are cool. Stacking greens and wreaths on warm ground or a dry

floor is asking for trouble. We pile our stored wreaths close together on the ground, preferably at night when the ground is cold. Don't stack them more than one bundle (5 wreaths) high, and pile them close together, so they will trap the coolness of the earth and hold it, even when the air becomes warm. If the ground has not yet thoroughly cooled in the fall, we pile them in a protected place and wait until a very chilly night; then we repile them on the frosty earth the next morning. Covering them with plastic is not a good idea, because on warm days it attracts and retains too much heat.

Decorating Wreaths

The addition of a bow and a few cones and berries can more than double the price of your wreath if you sell wholesale, and if you retail, the percentage of mark-up will be even greater. Tastes in wreath decorations vary a great deal, so it is wise to check what your buyer has in mind before you decorate. Some like only a large red or gold ribbon; others prefer only natural decorations, such as cones, pods, nuts, dried flowers, or the berries of mountain ash, bittersweet, cranberry, cotoneaster, holly, or viburnum. Or you may find that your clients prefer fruits, berries, Santas, and balls made of plastic, covered with artificial snow, sprays of gold and silver, or even a string of miniature lights operated with a battery.

Even though you might prefer hanging something different on your own front door, keep in mind Dale Carnegie's advice: "When you want

Wreath-Making Tips

- If you sell your wreaths retail, don't hesitate to test your artistic talents and use a variety of greens, and perhaps two or more kinds in combination. These can be especially appealing to those who are looking for something different from the wreaths that hang on their neighbors' doors.

- Whether you start around the wreath clockwise or counterclockwise, always do it the same way each time so it will become a habit. Consistency will add greatly to your speed and soon you will make each movement automatically without thinking about it.

- As you work, vary the size of the bunches if you notice that the wreath is becoming too heavy or light. If it appears floppy, you may be placing the bunches too far apart on the ring or you may be wiring too close to the cut ends of the twigs. If the wreath is not shaping well into a round form, use shorter lengths of greens.

- A 10-inch ring is probably the smallest size you should try to use with Scotch pine greens. Eight-inch ones are best made with firs and club moss, but even when using these greens it may be difficult to keep a center opening. Use short lengths and vary the size of the bunch, so the wreath will be full but still retain a flat appearance and a center. These small wreaths make nice bases for candles and other centerpieces.

- Instead of reaching about for wire cutters, or dulling your clippers by cutting the winding wire when the wreath is finished, make a kink in the wire where you want it to break and snap it suddenly. Voila — ten seconds saved!

- If you have some greens that are less than perfect, don't throw them away unless they are really bad. On a double-faced wreath you can wire a piece at the back of a bunch now and then, and if done carefully it won't show, but will add bulk to the wreath.

- When making different sizes of wreaths, I find it practical to make a few of each size every day. The longer and coarser greens can be used for larger ones, and the short pieces for the tiny ones. Flipping large double wreaths back and forth can be quite tiring on the shoulders, so alternating them with the smaller ones will make your muscles ache less that night.

- If you are filling orders for specific amounts of wreaths, keep a chart on the wall to mark the progress of each order, or count out the rings ahead of time so you will not accidentally make too many of the wrong size.

- Tie each bundle as soon as it is finished and immediately move it to a cool, shady place out of the wind.

to catch fish, bait your hook to please the fish." If you want to make sales, furnish what your market wants.

Wreath wire may be used to attach the decorations to the wreaths, but florists' picks are much faster and they can be attached to the decorations in advance of the wreath-making season. The green-painted-picks have a wire attached to which the cone, berry, or whatever is fastened; then the pick is inserted into the wreath. Precut florist wire is also available, which is a big help in attaching bows quickly.

Supply houses offer various kinds of fabric and plastic ribbon in different widths for making bows. Weatherproof types should always be selected for wreaths that will be hung outdoors. Making bows is difficult for some people, but ready-made bows and even bow-making machines are available (see Appendix).

Other Evergreen Decorations

Creating roping, or garlands, as they are sometimes called, is another popular way to make money from greens. Each holiday season many yards of roping are hung around windows, door frames, and porches, spiraled around lampposts, and draped over city streets. Roping is made either by hand or machine in much the same way as wreaths. Bunches of evergreen twigs are wired around a heavy twine, first on one side, and then the other. Nearly any kind of greens can be used, including spruce, if it is for outside use. A mixture of pine and fir makes especially attractive roping. It is easier to wire the greens to the twine if you suspend the twine from a ceiling hook and pull it down as the roping is completed. Roping is commonly made and sold in coils of 25 feet, or it can be made to suit your customers' requirements. It is tied in the same way that you would tie a stack of wreaths.

When you take an order for roping, find out what its use will be. Home use usually requires a light, tight roping made of short pieces of greens. For commercial use or street decoration it must be fashioned from larger pieces so it is more massive. Christmas greenery is, of course, not limited to wreaths and roping. Evergreens are also made into baskets, stars, crosses, hearts, trees, and doorway sprays. Wire shapes are available from supply houses or you can make your own from wood or rigid foam blocks. Corsages of evergreen twigs with a tiny hemlock or larch cone and a bit of red ribbon or berry are popular. So are evergreen centerpieces made on rings, pre-made florist's forms, or floral foam (such as Oasis), and these are often combined with small, white birch logs, driftwood, candles, or other objects. All can be offered retail if you have a shop, or sold wholesale to florists, garden centers, roadside stands, or Christmas tree retail outlets. You may even find a market for small, plastic bags full of evergreen tips that can be used for a variety of indoor decorations or stuck into window boxes for outdoor winter color. Some enterprising growers make up kits of evergreens, ring, wire, ribbon, and cones, complete with instructions for a do-it-yourself wreath.

Although tradition is "in" for the holidays, new ideas are always welcome, too. One of our artistic friends, who operates a small florist's shop, buys the December issues of several home and women's magazines each year, and during November, digs them out to study for creative ideas.

Working with greens can be an enjoyable and creative hobby. It is also an enterprise that produces color and beauty in those short and often dreary days when the brightness of flowers and the green of summer have disappeared. Making wreaths, garlands, sprays, and all the other decorations can also become a profitable business that can be started with little equipment and investment.

III

HARVESTING AND SELLING CHRISTMAS TREES, WREATHS, AND GREENS

If Christmas tree planting, shearing, and other cultural methods have been done perfectly for years, but the final operations — harvesting and selling — go wrong, all the time and work will have been wasted. Yet, sadly, it is at this point that a great many producers fail.

Some things about the business are out of our control and hard to anticipate. Markets can become over supplied, trees in a certain area may become diseased and quarantined at the last minute, extremely dry, warm weather or heavy snows can hit during the harvesting season, or truckers may go on strike. Although these things give nightmares to tree farm managers, sometimes you can alter your plans when things go wrong. There may be years, for example, when it is better to cut the trees a bit early or late, or even not to cut at all, and hope for better conditions the following year.

CHAPTER SIXTEEN

Marketing Trees, Wreaths, and Greens

Most aspects of market-ing are within our control, and by planning ahead it is possible to harvest successfully and sell all the good trees we are able to produce year after year. Long before your trees reach marketable size, for ex-ample, you must decide whether to sell them retail, wholesale, or both. If your production is large, the choice is likely to be wholesale. Few areas have a population large enough to retail thousands of trees at one location, but the huge wholesale market can include much of the world.

Wholesale versus Retail

Wholesale selling can take many forms. The easiest, even though it's at the bottom of the merchandising ladder as far as revenue is concerned, is to sell the trees "on the stump": The buyer selects the ones he or she wants, then cuts, wraps, and takes them away. This method of selling is recommended only as a last resort for

several reasons. Not only are the prices you receive much lower, but such buyers cut only the best trees, and often damage the others. And they usually leave a mess in the field, as well. Unless you work with the buyer, you must accept his or her count, also, so there is ample oppor-tunity to be fleeced on the number taken.

Cutting and wrapping the trees yourself is one step up. The prices are better, as is your control over the har-vesting. Delivering the trees to your buyers may also work to your advan-tage, but check out the buyer care-fully. Some have been known to try to negotiate the price downward once the shipment is in their yard.

Large producers usually sell their trees to a jobber who subsequently wholesales them to nurseries, florists, shopping centers, or filling stations, or to seasonal retailers who sell on vacant lots or from the back of a truck. Some growers prefer to omit the

117

jobber and sell their trees directly to these retailers. This sales method is another step up the selling chain in profit-making, but with each step upward, more responsibility, labor, and time are involved, and often the job of collecting the money from many customers becomes more difficult.

No matter what type of wholesale selling you do, it is important to line up your buyers early each year, even if you aren't sure of the number of trees you will have and must estimate on the low side. Most sales are made before July and many growers make their sales contracts even a year ahead. Anyone who cuts thousands of trees, piles them beside the road, and expects a buyer with a wad of money to appear suddenly in November is probably not going to spend the winter in Bermuda. And you can't be sure the buyer you have supplied for many years will let you know if he or she decides to get trees from another state, either, so it is important to insist on a deposit every time.

Fortunately, there are more sales aids for growers now than there once were. Marketing services are available through extension services, Christmas tree organizations, and cooperatives, but even with these aids most of the responsibility for selling lies squarely on the grower.

Retailing directly to the customer is the top of the sales ladder as far as profit goes. The price per tree may more than triple over selling it wholesale, but the risks also greatly increase. Collecting the money is not usually a problem, but weather, competition, and other factors may leave you with many unsold trees on Boxing Day, and no permit for a New Year's Day bonfire.

Grading and Tagging Trees

After the trees have made their early summer growth, tag the ones you want to cut that fall. That way you can confirm your wholesale orders and be prepared to sell any extra you may have discovered that are ready to harvest. If you sell your trees retail, you'll want to know in advance how many of each size you will have, to help you know how much to advertise and whether you will need to buy some elsewhere to provide a good assortment for your customers.

Check with your state Christmas tree association, county forester, or, in Canada, Agriculture Canada to see what method of grading is being used in your area, because methods vary from year to year and from region to region.

The following factors determine the quality of a Christmas tree: its general appearance; the degree to which its form is well shaped and symmetrical; its taper; the color of its needles; and whether or not it appears healthy.

The United States Department of Agriculture has established a grading system for Christmas trees and a few states have adopted it or a similar system. Three grades are suggested. A tree must be nearly perfect to rate a *Premium* grade. A *Number one,* sometimes called a *Choice,* can have one less-than-perfect "face," but the other three should be of high quality;

such trees can be placed against a wall. A *Number Two,* sometimes called a *Standard* tree, can have either a slimmer or fatter taper than a Premium or Number One, and only two sides need have a good appearance. These are often called "corner trees." A *Cull,* of course, is a tree that doesn't measure up even to Number Two.

Although these grading standards were established many years ago, grading of trees is not enforced and is done mostly on a voluntary basis. The only grades actually used by most growers are "salable" or "unsalable," although trees of many different qualities are sold within the range of "salable."

Growers should concentrate on growing only high-quality trees, of course. They are the ones that bring you the most satisfaction and the highest profits.

If you sell more than one grade of tree, grade them accurately, marking them with the various colored plastic ribbon (called flagging) available from supply houses. For the sake of your reputation, don't go easy on the grading. No one wants to buy an ordinary tree with a premium price. Keep a precise tally of the count. I've learned never to try to make an estimate, because during a casual walk through a planting most trees have a way of looking very good. It is only when you begin to put tags on the premium specimens that the number shrinks considerably.

Check the trees again throughout the season to make sure that disease or insects are not damaging them

and that the color stays good.

Trees are also often tagged according to their size as well as grade, and the National Color Code recommends the following:

2-3 feet: Blue
3-4 feet: Salmon
4-5 feet: Orange
5-6 feet: Red
6-7 feet: Yellow
7-8 feet: Green
8-9 feet: White

Correcting Poor Tree Color

Buyers will not be interested in your trees unless they look healthy. Trees that have been green all spring and summer sometimes mysteriously turn yellow in late fall or early winter even when there are no insects or diseases present. This condition may be the result of natural fall coloring, as is common with some strains of Scotch pine. It can also be a result of nutrient deficiency when a tree runs short of food just as it should be storing it up for winter. A rainy summer may leach away the fertilizer, or heavy weed growth in late summer may also rob the trees of nutrients. If you notice that the condition appears most autumns as the trees grow larger, apply a bit of extra fertilizer the spring of the same year the trees are to be cut. The color should improve by harvest time.

If fertilizing doesn't help, it may be necessary to spray a tree colorant a few weeks before harvest. These green dyes are especially formulated to be absorbed into the needles of evergreens. They look natural and will not wash off. Colorants are available from most Christmas tree suppliers (see Appendix) and can be applied with almost any typé of sprayer.

Dealing With Wholesale Buyers of Trees and Wreaths

Nearly every long-time Christmas tree or wreath producer can hold you enthralled for hours recalling experiences he has had dealing with buyers. Most of us have felt at one time or another that old pirates must have been reincarnated and gone into the business. But, in fairness, dealers also have some grim stories about growers who have reneged on agreements or shipped out products other than those specified in the contract. Christmas tree organizations, through their code of ethics, have in recent years succeeded in improving things greatly for both producers and buyers.

Even though conditions are better now, there are still buyers who should be cracking rocks on Devil's Island and growers who slip in trees that even Charlie Brown wouldn't buy. Both hurt the whole industry. We once had a buyer give us a deposit and then pick up part of the order, giving the truck driver just enough money to pay for that, but no deposit for the rest. He promised to pick up the rest of the trees the following week. That week never came, and we were left with piles of unsold trees. It is a common practice for some whole-

salers to order from several different sources, then collect only what they can sell or what is most convenient.

Some sellers refuse to start loading a truck before they see the money. Never assume that just because the check you got as a deposit was good that the final payment will be good, also. Bad checks are so numerous in the Christmas tree business that most transactions are made in cash or by certified check. Even some cash transactions are risky. We once caught a man counting out twenty dollars bills, some of which were folded over to look like two!

Try hard to find trustworthy buyers, and then work equally hard to keep them as customers by delivering what you promise every year. Whether you sell trees, greens, or wreaths, producing them is challenging enough without the extra task of locating a new market annually.

If you intend to deliver your trees or wreaths, be certain that trucks will be available when you need them. In some areas, pulp and lumber truckers also handle trees and greens, and in others, livestock trucks are available. If you expect to be involved in international sales, you will probably need to hire a customs broker to handle export and import regulations so the trees will arrive on time.

Even when you trust your buyer completely, get a written contract specifying every detail: the number and sizes of trees or wreaths, the pounds of greens, the date of pickup, and the prices agreed upon. If you don't know the buyer, try to get the names of other local growers with whom he or she has done business.

Always insist on a deposit. It is a common practice, so no one should feel insulted. A quarter or more of the total price is the amount usually collected, but it can be negotiated to whatever is agreeable to you both. When a buyer is unknown to you and provides no references, the deposit should be on the high side. Your bank can help you check out large established buyers through Dun and Bradstreet.

As in any business, there are regulations about selling trees and greens. You will need to provide your buyer with a sales slip or other proof of sale, and, if shipping out of the state, possibly a state and federal inspection certificate. Ask your county forester about this. If you sell living trees, either in pots or with the roots wrapped in burlap or plastic, you will also need a nursery inspection certificate from your State Department of Agriculture.

I know of loads of trees that have been confiscated because the trucks carrying them violated interstate commerce laws. If you deliver, therefore, be sure the trucks you use are properly registered to operate in whatever states they will be moving and that the owners are familiar with all regulations that apply.

Retail Selling

There are unlimited ways to market trees and wreaths directly to the consumer. For instance, you can sell from your front lawn or tree lot, rent a corner of the parking lot at a shopping center, take a load to a turnout area alongside a highway (if it is legal in your state), hire a vacant lot in a city, advertise through newspapers and magazines and ship directly to your customer, or manage a "choose-and-cut" operation and let your customers do the harvesting for you. One woman who lives on a little-used dirt road near us hangs a few wreaths outside her home each year during deer-hunting season. She claims that the hunters buy all she can make. They take them home as peace offerings to their neglected wives.

On your lot, wherever it is, stand your trees neatly on end by leaning them against a rack rather than having them thrown in a heap. This not only makes selection easier, but trees are more attractive when they are not crushed in a pile. Prices should be prominently displayed on weatherproof tags. To create customer goodwill, give your buyers a printed sheet with directions for caring for the tree. (See sample.) It will help you both.

A recent survey showed that some consumers buy artificial trees because they don't enjoy roaming around tree lots on cold windy days or nights, or transporting home a frozen, snow-covered tree. To encourage customers, make everything as comfortable and easy for them as possible. The nature of the business means you must be outdoors so the trees will be in a cool environment, but you can plan the layout of your selling area so that it is sheltered from the wind, and parking and pickup are easy. Provide windbreaks and shelters and perhaps even hot chocolate and coffee if your clients come

Caring for a Real Christmas Tree

Now that you have selected your special tree, please care for it as you would a fresh bouquet of flowers. Before setting up your tree, make a fresh, straight cut across the base of the trunk. About 1 inch is enough. Immediately place your tree in a water-holding stand, or, if you are not ready to decorate it, in a bucket of water. Don't ever let the base of the tree dry out or a seal will form and a new cut will be necessary. Check the water level daily. A fresh tree may absorb several pints to a gallon of water each day.

A Christmas tree, like a bouquet of flowers, likes it cool and safe, so please don't put it near your fireplace, heat source, or television set.

Don't forget to unplug lights when you go to bed or leave home.

Before decorating, be sure that the light cords and connections you use on your tree are in good working order to insure a safe Christmas for your family.

After the holidays a Christmas tree has many uses:

1. Place the Christmas tree in the garden or backyard and use it as a bird feeder. Orange slices will attract the birds and they can sit in the branches for shelter.

2. A Christmas tree is biodegradable; its branches may be removed and used as mulch in the garden. The trunk can be used for fuel or chopped for mulch.

3. Fir tree foliage can be stripped from the branches and snipped into small pieces for stuffing into aromatic fir needle pillows, for sofa or bedroom.

4. Large quantities of used trees make effective sand and soil erosion barriers, especially at beaches.

5. Sunk into fish ponds, trees make excellent refuge and feeding areas for young fish.

6. Woodworking hobbyists can make a multitude of items, including buttons, gavels, and candlesticks, from the trunk of a recycled Christmas tree.

Courtesy of the National Christmas Tree Association, 611 East Wells Street, Milwaukee, WI 53202; 414-276-6410.

from long distances. Try to make the checkout fast and convenient so chilly customers won't be kept waiting. Provide helpers to answer questions, carry the trees, and help tie them to cars. If you have only a dozen or so trees to sell, these won't be major considerations, but with a large opera-tion, they can be vital. Unfortunately, sales must all take place within a few weeks at one of the coldest times of the year, so be ready to use any legal means to lure customers into your lot instead of someone else's or to prevent them from driving to the mall for a fake tree.

If you want to reach working people, it will be necessary to be open nights and weekends and to arrange good lighting for the lot and your advertising signs. Don't overlook liability insurance when you are dealing with the public. Your agent should be able to write a policy that will cover you only during the weeks you will be in operation.

Selling Related Items

In addition to trees and wreaths, many retailers offer other enticements to attract people to their lots. Some sell tree stands and removal bags that double as tree skirts during the holidays. They sell wreaths, sprays, roping, and centerpieces so their customers can do one-stop shopping for their holiday decorating. For do-it-yourself decorators they also merchandise loose greens, wreath rings, wire, cones, berries, and ribbon, as well as candles, artificial snow, poinsettias, and lights. Imaginative merchandising pays off and displays that show how the various products can be used always stimulate more sales: a lighted, decorated tree, wreaths hanging on doors, roping around a window or lamppost, or a centerpiece on a table or mantel of an artificial fireplace. Some sellers, taking their cue from department stores no doubt, use tapes of recorded carols to help buyers get into the right mood.

Choose-and-Cut Operations

Like pick-your-own berries, cut-your-own trees has become a popular way of merchandising. Tramping through a plantation is much like going out in the wilderness to cut a tree, and families with children particularly enjoy making this ritual part of their holiday tradition. Many bring their cameras and may even include an outdoor winter picnic. People also like the lower price and fun of searching for the perfect specimen from a large selection of farm-fresh, well-spaced evergreens.

To be successful, a choose-and-cut operation must be located where there are enough customers within driving distance to match the number of trees available. The planting should be accessible to the road, easy to find, and include plenty of parking space located so that customers won't need to carry their trees a long distance. Some growers plant blocks of trees in consecutive years along roadsides, so that each year a new block will be ready near a parking area.

Others prefer to plant new trees each spring where the mature ones were cut the previous fall. Such plantings are appealing because the trees are more natural looking, with all sizes growing together. Unfortunately, in such lots the customers sometimes inadvertently damage small trees, especially when the trees are partially snow covered.

You will have to develop your own method of pricing. The simplest way is to ask the same price for every tree. Some tag trees with different colored flagging, or paint various colors to represent the different prices on the tree trunk just above where it will be cut. It was once a common practice for growers to let customers tag trees early in the fall for later cutting. Most growers no longer allow this, because tag switching was common, as was the inescapable feeling that the best trees had already been reserved. Other problems included the no-shows and lost souls who needed help to find their tagged trees. If you allow pre-tagging,it is best to set a cut-off date for such trees to be picked up.

Not much harvesting equipment is needed for this type of operation, although you must provide small bow saws or pruning saws for your clients. Axes are too dangerous for novices, and they leave a sharp stump. Since customers are often dressed in clothing that is not warm enough and may take awhile to choose their tree, a warming shelter and hot drink is an especially welcome addition to a cut-your-own lot. Also, you may want to have a salesperson standing by to help customers shake the old needles out of the tree if pines are sold, and to tie the tree on the car. Extra services like these will help make the tree cutting a pleasant experience your customers will want to repeat.

Some choose-and-cut lot owners even provide rides to distant fields in horse-drawn sleds complete with sleigh bells or in a trailer pulled behind a tractor. Such junkets are fun, but because they take extra time during a busy season, the costs of such additional services must be added to the price of the trees. Because of the higher accident risk, insurance costs will likely be more, too.

One big plus for a choose-and-cut operation is that if business hap-pens to be bad any year, the unsold trees don't have to be destroyed but can remain growing in the field, available for the next Christmas season.

Pricing

Deciding how much to charge for trees, wreaths, and greens is difficult, even for old-timers, and newcomers often find it frustrating to set a fair price for each size and grade. Unfortunately, in both retailing and wholesaling, the prices charged by the competition usually determine the price you can ask. Check with the marketing department of your extension service for information about current pricing in the industry. Those selling wholesale often depend on their buyers to set the price, but this isn't usually the best deal for the producer. Sometimes after a long session of haggling, the seller gives up and takes what he can get.

It is important, in any case, long before they are sold, to know what your production costs have been and how much you must charge to make a fair profit. Try to be reasonable. Cutting prices to get business hurts the entire industry and inflating them when trees or greens are scarce speeds up the sales of artificial trees.

Discounting trees or wreaths the days just before Christmas is one way to get rid of them, but it presents a dilemma. You can't be sure how many customers will appear during those last hours, yet you don't want to get caught with leftovers. Adding to the problem is the fact that many customers expect trees to be marked down at the last minute, so they wait until

---

Free or Low-Cost Advertising

- Many towns in rural areas have bulletin boards in public places where notices of temporary interest can be posted.
- A magnetic sign on your truck or car draws attention to your business, especially in a rural area. I learned how effective these are one time when a policeman stopped me on the highway. Instead of a ticket, though, to my great relief, he wanted to order a tree!
- Newspapers, as well as radio and television stations, often feature local industries, particularly those of seasonal interest, and may be happy to do a feature on your Christmas trees in early December. If this happens, prepare a fact sheet for them in advance, both to save your time and in the interest of accuracy.
- Community service projects benefit both you and the community; for instance, a local institution gets a free tree to use or auction off and you receive free publicity. Donate a few trees, wreaths, or other decorations to local schools, parks, churches, or clubs. Furnish centerpieces for local luncheons. Sponsor classes on how to make Christmas decorations. Make up instruction sheets your students can take home (with your business name on them, of course). You can probably think of numerous other opportunities that would be tailor-made for your community.
- Last, but far from least, don't miss any opportunity to promote good will and customer satisfaction. Make sure your salespeople realize that to your clients, they are the business, and to a great extent, their manners will determine whether the buyers will recommend you to their friends. Loyal customers are your most effective, and least expensive, advertising.

---

then to shop.

People who haggle over price can be expected to appear occasionally throughout the season, a practice that is likely to accelerate on the last days before Christmas, so be prepared to handle it. There are always cases when you should be willing to give a special customer a break, but it is best not to get a reputation for being an easy mark, unless you prefer to bargain for every tree.

After you have determined your prices, be sure each tree is clearly marked. Marking allows your customers to serve themselves without bothering you and discourages bargaining. Unmarked merchandise gives the impression that a different price is made up for each customer and that doesn't bode well for your business.

Retail Advertising

No matter how successful your business, some sort of advertising is necessary each year to remind customers that it is time to pick up their trees. If you are new at retailing, you will, of course, need advertising to introduce your business. On-the-lot advertising can be done with attractive signs or vinyl banners and pennants (available from supply catalogs). An attractive year-round sign at your plantation is helpful, too, if you intend to sell trees in that area. It may seem silly to advertise Christmas trees in May, but if your sign says "Whittakers' Christmas Tree Farm," passersby will be likely to think of you when sleigh bells ring in December.

Off-site ads are vital if you are not located in a heavily travelled area. Next to pricing, deciding how to spend the advertising dollar is one of the most difficult decisions a business person must make. There are so many radio and television stations and newspapers in most regions that you probably have a wide choice and should check out the various possibilities carefully. Direct-mail advertising is also a good sales technique, but unless you have your own mailing list this method is probably too expensive to be practical.

Mail-order Selling

You can operate a successful retail business for trees, wreaths, greens, and other products, even if you live far from a populated area by selling by mail. The post office and private carriers such as United Parcel Service make the whole country your marketplace, and a few ads in carefully chosen magazines can bring you a host of orders. After your business has been around for a few years you will have developed a mailing list to send out brochures or circulars.

Make sure you are able to handle the business you get. Orders must be handled rapidly and accurately, so the facilities and equipment for packaging, labeling, and bookkeeping must be well organized far ahead of time. It may be possible for you to share the equipment of a nearby mail-order firm — scales, wrapping, and so on — and you may even want them to do the merchandising for you along with their regular business. Follow any regulations about inspection and certification of trees or wreaths leaving your state. Price your trees and wreaths realistically. Your expenses will be much higher than if you were selling them off a lot.

CHAPTER SEVENTEEN

The Harvesting

Like most farm operations, tree cutting must be done at precisely the right time, but unlike crops such as strawberries, grapes, and asparagus, Christmas trees cannot be processed, stored, and sold at a later date. Because they must be harvested, sold, used, and disposed of within a period of only a few weeks, this calls for careful planning, hard work, and sometimes just plain luck.

Trees and greens should not be cut until their new growth has been hardened by a few days of cold weather, and until the danger of prolonged periods of hot, dry weather is over. The season may begin as early as mid-October in the northern mountains, but a month or more later in warmer regions. As I've said before, tree growers, unlike the rest of the population, always hope that cool, moist weather will arrive early and that Indian summer will be short.

Anyone who tries to lengthen the harvest period by cutting a few weeks early usually regrets it. Many growers have forfeited future sales by furnishing buyers dried out trees and brush. Attempts to outwit nature by storing trees and greens in refrigerated houses and potato cellars have not proved too practical, either. One grower I know stood several hundred trees upright in a shallow pond on his farm one fall. He was glowing with pride as he admired their rich green color a month later. Unfortunately, his buyer wanted to pick them up the last week of November and winter came early that year. The day the trucks arrived, the trees were frozen in 3 inches of ice and the ice was covered with 8 inches of snow.

Some people spray their trees and wreaths with an antitranspirant such as Clear Spray to prevent them from drying out, but the best way to keep trees fresh, if you can manage it, is to cut them as late as possible and pray that cold weather lasts until Christmas.

Selecting Harvesting Equipment

Because it will be several years after planting before you need harvesting equipment, this purchase can be delayed for a while, and this will give you time to select what is best for your operation. Since the season is so short and speed is crucial, power equipment is essential for all but the smallest operations. Be sure to have everything ready in advance — wrapping machinery, wrapping materials, and sharp saws that are in perfect running condition, with spares of each vital tool on hand, just in case.

For most operations, lightweight chain saws are the best choice for harvesting. Even on small operations they are practical because they are faster and need less muscle power than bow saws, although the latter usually are adequate and safer. Bow saws are low priced, trouble free, easy to carry around, and as a testimonial, they are popular with tree thieves, who try to keep the noise level down as they work. Large growers need power saws of some type. The shoulder-carried machines with a circular saw at the end of a shaft are easy to carry and use, but chain saws are more popular. On very large plantations where most of the trees on the lot will be harvested at one time, the heavy-duty, wheel-mounted circular saws are used because they are most efficient and fastest.

With all power saws you must be extremely careful. Not only is it easy to cut yourself, especially when you get tired toward the end of a long day, but, because you are cutting close to the ground, it is also easy to saw into the ground or a rock and ruin your machine. Wear a face shield, or goggles, chaps, and possibly a hard hat for protection.

Since the trees have probably been basal pruned (see page 74), the cutting can go rapidly. The "handle" on the cut tree should be approximately 12 inches long on a 7-foot tree so the buyer can saw it off once more before mounting it in a tree stand. Some experts cut 1,500 or more tagged trees a day.

One of the more muscular jobs in harvesting comes in getting the trees from the lot to the loading area, and it may take several workers to keep up with one good cutter. Lightweight trees can be carried by hand and heavy ones can be safely dragged over snow-covered ground, but never pull large trees long distances over rough, bare ground. Such treatment is likely to scrub off bark and needles and ruin their appearance — another good reason for having plenty of access roads.

Plan your operation so tree handling can be kept at a minimum. Handling costs money and can be hard on the trees, so the closer the truck can drive to the growing area, the better. At least one large grower uses a helicopter to move trees from a back lot to the loading area, but most of us don't think in terms that big.

For a small, retail-type operation a tree-tying or wrapping machine isn't necessary, but will be essential for a wholesale operation. Loose trees not only take up a great deal of room on a truck but are easily damaged in handling, especially when frozen. They also dry out more quickly. Most buyers require that trees be wrapped, although some insist on doing the job themselves, so they can inspect each tree first. Some gasoline-powered machines spin the tree while wrapping it in baler twine, but a more modern gadget, operated entirely by hand, wraps the tree in a tough plastic netting. With the latter, a tree is slid,

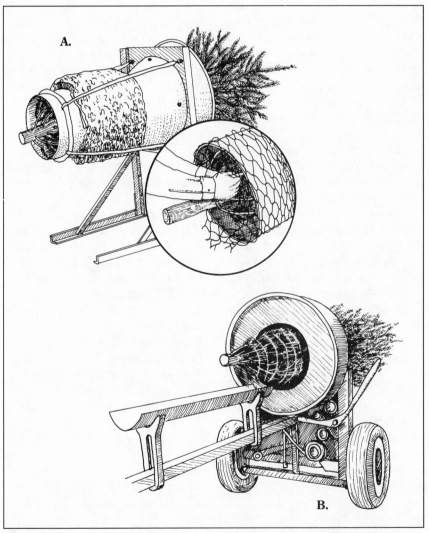

(A) *A tree is wrapped in plastic netting by being pulled, butt-end first, from a sleeve that dispenses the netting.* (B) *A gasoline-powered machine spins the tree while wrapping it in baler twine.*

butt first, through a cone of the right size, picking up the plastic wrapping from a sleeve as it emerges from the cone.

The initial cost of the power-operated string wrappers is more than those using netting, but the twine costs far less than the plastic netting. Before you invest in tree-tying equipment, find out how your buyers are likely to want the trees wrapped. Some retailers prefer the neat-looking plastic netting-wrapped ones to those tied in string, but others complain about the difficulty of removing and disposing of the plastic.

Before wrapping, pines should have their dead needles removed by thumping them on a solid surface, or with a mechanical tree shaker.

If yours is a wholesale operation, you may also need a conveyer to lift the trees to the top of a large truck. Again you may not need to buy one. Dairy farmers do not use their hay bale conveyers in early winter so some tree producers rent these for the tree harvesting season.

Other Harvesting Precautions

If the weather is dry for several days after cutting, it may be necessary to sprinkle the trees with water to prevent them from drying out. It is best to do this at night or early in the morning, because water poured on warm branches evaporates quickly and the steamy moisture is likely to loosen the needles. A small number of trees may be hosed down, but sprinklers such as those used in automatic irrigation systems will be necessary for large operations.

In some parts of the country, customers like trees that have been "flocked" or sprayed with an artificial snow. Although most wholesale buyers prefer to do this themselves, some growers do it on their farms. Flocking machines and materials are available from supply houses (see Appendix).

If you hire a trucker, try to get one who is accustomed to loading and hauling trees. An experienced packer can get two or three times as many trees on one load as a novice who just piles them on haphazardly. Put the smaller trees on the bottom where they will pack tightly together and save the larger, heavier trees for the top. Walk over the load when packing to compress the trees, being especially careful if they are frozen, because frozen trees are brittle. Pile the trees as high as is practical, being careful to keep within the legal trucking weight and height limits. Round off the load by piling the last trees only in the middle. Then tie them down securely. Check the load a few miles down the road. Even if they were loaded skillfully, the trees will settle and need retying.

CHAPTER EIGHTEEN

A Checkbook with a Happy Ending

There are Christmas tree growers who know how to raise trees, and there are those who know how to run a business successfully. The lucky ones know how to do both. If you are good at only one, however, you may want to hire someone else to help with the other phase of your business. A lot of successful plantations are owned by people who are smart enough to hire excellent managers to run certain aspects of the operation.

The Organized Bookkeeper

Before you set up your bookkeeping system, take a look at the income tax forms you will eventually be filling out, so you can keep your deductible expenses separate from those that are not deductible. At the present time, if you operate a farm, you can figure your Christmas tree and greens business as part of your farming operation and report everything on Schedule F. If your planting is not part of a working farm, however, you must report everything on Schedule C. For some unknown reason, the IRS does not consider trees an agricultural crop, just as they do not list raising fish or earthworms under farming.

As you set up your record system, keep it as simple as possible or you'll be unlikely to do a good job, especially during your busy seasons. Record day-by-day memoranda in a pocket notebook and one day each week transcribe your notes into a permanent record book.

Good bookkeeping not only helps you fill out tax forms more easily, but also could help you to deal better with an audit if you ever have to face one. It can also guide you when setting up a workable budget for future years and warn you when costs are getting out of line.

Operating and Capital Expenses

Your expenses will be of two types — either operating or capital. Operating costs can be deducted during the year that they occur, but most capital expenses must be depreciated over a period of time.

Operating costs include, among other things, office supplies, organization dues, fuel, repairs, upkeep, legal and consulting fees, business travel, advertising, land rent, fertilizer, pesticides, and machine hire. Hired labor can also be deducted, except any used for capital improvements, such as work on new buildings. You can hire your own children and deduct their wages, but there is a limit to the amount you can pay them and still be able to deduct them as dependents, and before they have to pay taxes on their earnings. Capital expenses, the long-term costs, include land, seedlings, transplants, planting costs, buildings, machinery, and other equipment.

We manage our operation on a five-year plan. It takes longer than that from seed to sale, but even the best-laid plans need revising at least every five years. It is worthwhile to review periodically what is happening and take stock. How many of the trees planted seem to be making it to market? How does income compare with cost at this point? (You may want to evaluate this one two ways — both with and without your own labor.)

In a project that is small to medium in size, a calculator, ledger, and blank record book are probably all the bookkeeping supplies you need.

If your operation is medium to large, however, you will find a computer as helpful as a bookkeeper. It can write your checks, keep your accounts, figure your withholdings, payrolls, and taxes, as well as keep track of your inventory, machinery maintenance, and labor. Even before you start your operation, working with a simple spreadsheet can help you try out various "what if" situations, so that potential mistakes are made on a floppy disk instead of in the field.

No matter the size of your business, each winter make up a calendar similar to the one on pages 135-36, but adapted to your own business. There are many things that must be done each season and it is easy to forget some of them unless you have a reminder.

You and Your Government

Getting along with the government is an integral part of any business these days. Whether your operation is in the United States or Canada, it is important to know all the regulations that affect your business, because, as we are so often told, ignorance of the law is no excuse. Each time lawmakers meet, I notice cynically that they seem to make life more complicated for business people, and usually more expensive. In the U.S., as soon as you start your operation Uncle Sam will be looking over your shoulder, along with your state and local officials. Someone will want to know not only all about your expenses and income, but also about the chemicals you use and the dates you use them.

When you hire employees you will need to withhold federal, state, and local taxes, including social security and probably workman's compensation. Write the Internal Revenue Service and your state tax department for these forms. If you retail trees or any other products, you will need to register with your state tax department and collect sales taxes, unless you live in a state that doesn't have a sales tax. If you sell part of your trees and greens wholesale, or to a nonprofit organization, you will need to fill out an exemption form with the tax numbers of those buyers, making a notation on why they are exempt. These forms must also be filed regularly with your state tax department.

If you construct buildings or set up a retail selling area, or if the business will generate additional traffic, you will probably need a permit from a local zoning board and perhaps a planning or environmental board. If a new driveway is in your plans, you may also need official permission for a highway curb cut, as well as permits for drilling wells and sewage disposal. You may need a permit to put up a sign and to install outdoor lighting, even if it is on your own property. And, if your business has a name, that must be registered and recorded.

You may find it an advantage to incorporate if your business is large, but study this possibility carefully. Incorporation provides certain advantages in taxes and in handling investment income, and it makes it easier to transfer the business. On the other hand, it also involves much additional bookkeeping and more forms to mail to the government. A corporation is also more complicated to terminate than a personal business or a partnership, should you so desire.

Until 1987, when Christmas trees were cut and sold, they were considered capital gains instead of straight income, but the tax revision of that year ended this advantage. As this book is being written there is an attempt to get this policy reinstated, so you will want to keep up to date about this development as well as all other changes in the tax regulations.

Keep in mind that in a democracy you are an influential part of the government and can make a difference with the laws. Most regulations are made by urban lawyers and legislators who are unfamiliar with the grassroots of the Christmas tree industry. Write to them, attend public hearings, and support your Association's efforts to oppose and change any bad legislation.

Keeping Up to Date

It is important to keep abreast of what is happening in this constantly changing world. Not only do government regulations concerning taxes, chemicals, and quarantines frequently change, but so do cultural methods and the marketing situation. To keep up to date, join your area Christmas Tree Association and attend their meetings. Subscribe to Christmas tree newsletters and magazines, read the bulletins published by your extension service, and attend any meetings they sponsor that pertain to the growing and marketing of trees.

It is always risky to make predictions about the future, as evidenced by the fact that most of the pessimistic forecasts issued twenty years ago about the demise of the Christmas tree industry simply didn't take place. Now, as the emphasis shifts from more efficient production to better merchandising, it is even more important for all the producers to work together. If this happens, there is no reason why the Christmas tree business won't have a great future.

The lists of dos and don'ts may seem overwhelming at first, as in any new project, but reading about it makes it seem far more complicated than it is. If you've read the book, you now know far more about growing trees than most of us did when we started, and there's no reason why you, too, can't have great success. I hope you'll enjoy being involved in this great industry as much as we have.

CHAPTER NINETEEN

A Christmas Tree Grower's Calendar

Since each situation is different, this seasonal calendar can serve only as suggestions of jobs to include on your own annual reminder list. For example, the schedule for management of planted trees differs from that of a wild lot. Each job listed doesn't need to be done every year, of course, and you will add chores of your own that aren't included here.

Winter (After Christmas)
- Review your records and make plans for the new year, including lining up workers if you will need them.
- Take inventory of fertilizer, pesticides, small tools, and office supplies. Order what you'll need for this year.
- Order seedlings and/or transplants, office supplies, etc.
- Do accounts, taxes, including sales tax.
- Update mailing lists of retail or wholesale customers.
- If the snow isn't too deep, basal prune trees and do any necessary corrective shaping of firs and spruces.
- Remove culls and clear brush from plantings and roads.
- Clean sheds and other buildings and do any remodeling.
- Get machinery in shape.
- Attend winter meeting of Christmas tree organizations.

Early Spring
- Apply preemergent herbicides, dormant sprays.
- Prepare transplant beds.
- Plant seedlings and transplants.
- Fertilize established trees.

Late Spring
- Cultivate, mow, or spray around trees.
- Watch for insect infestations and take necessary action.

Early Summer
- Shear trees and continue mowing.
- Spray, as necessary, to control dis-

ease and insects.
- Begin to line up wholesale buyers for fall.
- Do second (light) fertilizing of spruces and firs if they need it.

Mid- to Late Summer
- Check trees for aphids and other pests.
- Make preliminary inventory of salable trees.
- Order wreath rings, wire, tree tags, and flagging ribbon.
- Continue grass, weed, and brush control.
- Reinstate your liability insurance if your operation is retail or "choose and cut."
- Plan advertising, order signs.
- Visit other tree farms.

Fall to Late Fall
- Check trees frequently for disease and insect damage.
- Watch for poor color and spray trees with colorants, if necessary.
- Spread rodent poison.
- Allow rabbit and deer hunting if those animals are a problem.
- Get harvesting equipment in shape.

- Confirm the availability of your laborers for harvesting.
- Spread preemergent herbicides
- Spray Roundup for late summer brush control.
- Repair any road damage.
- Measure and tag trees for cutting and for choose-and- cut sales.
- Begin advertising.
- Cut greens as soon as the weather cools.
- List any unsold trees, greens and wreaths with the extension service or Christmas Tree Association sales bulletins.
- Put up your signs if they aren't up already.

November and December
- Take advantage of any free or low-cost publicity.
- Make wreaths and other decorations.
- Cut trees.
- Watch for poachers.
- Set up your retail stand and sell trees.
- Deposit money, pay bills, and enjoy the holidays.

Appendix

Christmas Tree Growers' Supplies

Abbott Laboratories
Chemical and Agricultural
   Products Division
1 Abbott Park Rd.
Abbott, IL 60064
*Growth regulators for white pine
   trees*

Aquatrols Corporation of
   America, Inc.
1432 Union Ave.
Pennauk, NJ 08110
*Folicote anti-dessicant; water-absorb-
   ent root slurries*

Bailey's
Highway 101, P.O. Box 550
Laytonville, CA 95454
Foresters' and loggers' equipment

Bluewater Import/Export
1627 Tyrie Drive East
Sarnia, Ontario N7V 3P6, Canada
Or P.O. Box 610846
Port Huron, MI 48061
Christmas tree stands

The Campbell Co. Inc.
P.O. Box 780
Wautoma, WI 54982
*Complete line of supplies and equip-
   ment for both growers and retailers*

CompuTree, Inc.
P.O. Box 4013
Englewood, CO 80155
*Computer software for Christmas tree
   growers*

Country Home Products
Box 89
Cedar Beach Road
Charlotte, VT 05445
Wheel-mounted string mowers

Electro-Spray Mfg. Inc.
P.O. Box 29335
Lincoln, NE 68529
Mist sprayers

Evergreen Sales and Mfg. Co.
P.O. Box 108
Birmingham, MI 48012
Wreath-making machines

Glamos Wire Products
5561 N. 152nd Street, P.O. Box 46
Hugo, MN 55038
Wreath wires and rings

A.F. Hillman Wreath Ring Co. Inc.
119 Dale Ave.
Patterson, NJ 07501
Wreath-making equipment

Howey Tree Baler Corp.
Merritt, MI 49667
Tree tyers, balers, elevators, and twine

International Forest Seed Co.
Blair Farm Road, P.O. Box 290
Odenville, AL 35120
*Tree seed, seedlings, and container-
    ized seedlings*

Kelco Industries
P.O. Box 160
Milbridge, ME 04658
*Planting augers, shearing tools, and
    wreath-making machines*

The Kirk Company
R.R. 3, Box 590
Wautoma, WI 54982
*Tree-tying equipment, colorants,
    needlehold, and balers*

A.M. Leonard, Inc.
P.O. Box 816
Piqua, OH 45356
*Pruning, spraying, labeling, and plant-
    ing equipment; supplies*

McCanse Engineering Services
949 Etnyre Terrace Rd.
Oregon, IL 61061
Tree shakers

Mechanical Transplanter Co.
1150 Central Ave., Box 1708
Holland, MI 49422-1708
Tree planting machines

Mitchell Metal Products, Inc.
P.O. Box 207
Merrill, WI 54452
Wreath rings of many different designs

North Star Evergreen
P.O. Box 253
Park Rapids, MN 56470
*Planting, shearing, spraying, flocking,
    and tying machines; chemicals,
    colorants, and wreath-making sup-
    plies*

Precision Steel Structures, Inc.
Rt. 15, Box 2498
Maryville, TN 37801
Portable Christmas tree sale huts

SAJE Inc.
6392 Portland Rd. NE
Salem, OR 97305
Power shearers; elevators

Sullivan Manufacturing and
    Sales Corp.
P.O. Box 666
Hammond, IN 46325
*Snow-flocking machines and
    supplies*

Utility Tool and Body Co. Inc.
P.O. Box 360
Clintonville, WI 54929
Tree planters

Veldsma and Sons, Inc.
P.O. Box 6
Forest Park, GA 30051
*General line of equipment and sup-
    plies*

Vermeer Mfg. Co.
Route 2, P.O. Box 200
Pella, IA 50219
Brush chippers

Yule Forest
1220 Millers Mill Road
Stockbridge, GA 30281
Power tree trimmers

Seed and Nursery Companies

Armintrout's
West Michigan Farms Inc.
1156 Lincoln Rd.
Allegan, MI 49010
*Spruce, Douglas fir, and pine
seedlings and transplants*
616-673-6627

Carino Nurseries
P.O. Box 538
Indiana, PA 15701
*Spruce, fir, and pine seedlings and
transplants*
412-463-3350

Drakes Crossing Nursery
19774 Grade Road SE
Silverton, OR 97381
Tree seedlings of western species
503-873-4932

Flickingers Nursery
Sagamore, PA 16250
*Wide variety of seedlings and
transplants*
412-783-6528

Georgia-Pacific Corporation
Mill Street
Woodland, ME 04694
*25 different strains and species of
conifers*
207-427-3311

Greener'n Ever Marketing
P.O. Box 222435
Carmel, CA 93922
*Bare-root and container seedlings;
growers' supplies*
408-624-2149

Lawyer's Nursery, Inc.
950 Highway 200 West
Plains, MT 59859
*Tree seeds, conservation plants, and
wildlife attracters*
406-826-3883

Roger Marcoux Plantations and
Nurseries, Ltd.
455 King St. E.
Sherbrooke, Ontario J1G 1B6,
Canada
Balsam transplants

Musser Forests
P.O. Box 340
Indiana, PA 15701
Tree seedlings
412-465-5685

Needlefast Evergreens, Inc.
4075 W. Hansen Road
Ludington, MI 49431
Spruce, fir, and pine seedlings
616-843-8524

Pound's Nursery Inc.
9835 Maharg Road NE
St. Louisville, OH 43071
Canaan fir seedling plugs
614-745-5946

F.W. Schumacher Co.
36 Spring Hill Road
Sandwich, MA 02563-1023
Forest tree and shrub seed
508-888-0659

Strathmeyer Forests, Inc.
255 Zeigler Road
Dover, PA 17315
Spruce, fir, and pine seedlings and
transplants

Walker's Tree Farms
Evergreen Acres
R.R. Box 616
Orleans, VT 05860
Spruce, fir, and pine seedlings and
transplants
802-854-8487

Western Maine Nurseries
P.O. Box 250
Fryeburg, ME 04037
Spruce, fir, and pine seedlings and
transplants
207-935-2161

Van Pines, Inc.
West Olive, MI 49460
Spruce, Douglas fir, and pine seedlings
and transplants

State and Regional Christmas Tree Associations

The officers of State and Regional Organizations change frequently, and the secretaries ordinarily operate from their homes. Contact your Extension Service for the name and address of the current secretary of the group in your area.

Alabama	Maine	New York
California	Maryland	North Carolina
Connecticut	Massachusetts	North Dakota
Florida	Michigan	Northwest
Georgia	Mid-South	Ohio
Illinois	Minnesota	Pennsylvania
Indiana	Missouri	Rocky Mountain
Inland Empire	Montana	South Carolina
Iowa	Nebraska	Texas
Kansas	New Hampshire-	Virginia
Louisiana-	Vermont	West Virginia
Mississippi	New Jersey	Wisconsin

Glossary

Asexual. The propagation of a plant by cuttings, layers, grafts, or other means rather than by seeds.

Bark. Protective outer covering of trees and shrubs.

Basal prune. Remove the lower branches of a tree, to create a stem for easier cutting and handling. On tall wild trees, basal pruning is often used to produce a salable tree high up on a stem.

Biodynamic. The use of organic methods to raise plants, as opposed to using chemicals.

Boughs and greens. The outer growth of evergreens, removed from the tree and used for making wreaths, roping, and other decorations.

Blight. A condition caused by fungus or bacteria, or a physiological situation that kills or adversely affects the growth of the plant.

Branch. Lateral stem of a tree.

Broadcast. To sow seeds, herbicides, or fertilizer by scattering.

Bud. The small swelling on a plant from which grows shoots, leaves, or flowers.

Budding. A type of grafting by which a single bud rather than a piece of branch is transplanted from one plant to another by surgery.

Cast. Hue or color, when referring to the tree's foliage.

Compost. Humus that is rich in nutrients; made from decaying plants, leaves, manure, and other organic matter.

Cone. The scaley fruit of conifers, which contains the seeds.

Conifer. A tree that produces cones.

Cull. An inferior tree.

Caliper. Diameter of the stump of a tree.

Cambium. The green layer between the bark and the sapwood of a tree where growth takes place.

Candle. The current season's growth of pines.

Choose-and-cut. A combination harvesting-marketing operation where customers select and cut their own Christmas trees.

Clay. A type of earth comprised of very fine rock particles that were once layered in water.

Climate. Long-term characteristics of weather for a region.

Conifer. A tree or shrub that bears its seeds in cones.

Cover crop. A crop, usually grass or a grain such as winter rye, buckwheat, or oats, raised to build up organic matter and fertility in the soil and to choke out weeds.

Damping-off. A general term for several fungus diseases that affect new seedlings.

Deciduous. Relating to plants that lose

141

their leaves over the winter.

Dendrology. Science dealing with trees and plants and their classification.

Ecology. The science of the mutual relationships of living organisms and their environment.

Evergreen. A plant that holds its foliage all year.

Fertilization of plants. The pollination of the flower.

Fertilization of soil. The adding of nutrients to the soil.

Fungicide. Material used to control fungus diseases.

Fungus. Disease organisms that attack plants.

Germination. The sprouting of seeds.

Graft. To transplant a portion of a tree or plant on to another by surgery.

Handle. The base of a Christmas tree's trunk.

Hardened off. The condition of a plant's new growth when it has become woody in late fall in preparation for winter.

Heel in. To plant temporarily.

Herbicide. Chemical used to destroy vegetation or to prevent the germination of seeds.

Humus. Partly decayed organic matter.

Hybrid. New plant (or animal) created by the crossing of two subjects that are different but within closely related species.

Insecticide. Substance used for the killing of insects.

Larvae. The immature, grub-like stage of certain insects.

Lateral branch. Side branch of a tree or plant.

Leach. Remove plant nutrients or salts by water percolation.

Leader. The top branch of a tree that grows straight up.

Lime. Naturally occurring mine ral containing calcium carbonate or magnesium carbonate used to make

soil less acid.

Loam. Rich soil containing a mixture of finely ground rock, clay, nutrients, and organic matter.

Manure. Organic fertilizer, usually bird or animal excrements, used to improve soil.

Marginal land. Land that is low in fertility, and unprofitable for ordinary farming.

Mulch. A covering for the soil, usually of organic matter such as hay, shavings, or similiar products, or inorganic substance such as crushed rock or plastic. Mulches are used for soil building, moisture retention, weed and erosion control, and appearance.

Nutrient. Mineral necessary for the growth of plants and animals.

Offshoot. A sprout growing from the main trunk near or just under the ground.

Organic. In farming, refers to the growing of plants by fertilizing with natural products such as rock phosphate, composts, and manures, rather than chemical nitrates, phosphates, and so forth; also the control of pests by using natural methods instead of chemicals.

Pathogen. Minute organism that causes disease.

Peat. Organic material that is partially decayed, found in bogs. Useful for adding humus and acidity to the soil.

pH. Method of measuring the acidity-alkalinity ratio in the soil. A pH of 7 is neutral. Most tree species do best on soils between 5.5 and 6.5.

Pollen. Male reproductive cells of a flower, borne on the stamens. Usually a fine, yellow or brown dust.

Pollination. The fertilization of the female portion of a flower by pollen from the male portion. Pollen is usually spread between evergreens by wind.

Prune. To cut away unwanted branches of a tree.

Root stock. A seedling tree upon which is grafted an improved variety of the same, or a closely related species.

Scalping. The scraping away of the upper layer of sod when hand planting to give young trees a better chance of survival.

Scion. A small twig used for grafting onto another tree.

Seedling. A tree grown from seed, rather than a cutting or some other method.

Selective cutting. Harvesting trees by selecting only certain ones, as opposed to clear cutting everything.

Shearing. The removal of the tips of branches to encourage a tree to grow into more compact shape.

Silviculture. The science of growing trees.

Soil structure. The general makeup of the various particles of mineral and organic matter in the soil.

Sport. A plant mutatation that shows definite differences from others of the same species.

Sprout. New growth on a plant.

Stratification. The storing of seeds at optimum conditions of moisture and temperature for a few weeks before planting to improve their germination capacity.

Stress. A weakening condition of a plant caused by extremes in weather, excess or shortage of nutrients or moisture, or similar circumstances.

Stump culture. The raising of a new tree from a branch left on the stump of a tree that has been harvested.

Subsoil. The layer of soil between the topsoil and the hardpan below.

Sucker. Shoot coming from the roots or lower stem of a plant.

Taproot. The main root of a tree.

Thin. Remove a part of the trees in a planting to give those remaining more room to grow.

Tillage. Land that is being cultivated for crops.

Tips. The ends of evergreen branches.

Tissue culture. The propagation of plants in laboratory test tubes under sterile conditions and carefully controlled temperatures.

Top-dress. Spread fertilizer over the ground without cultivation into the soil.

Topping. Removing the top of a plant, either to control the height or to encourage side branching.

Topsoil. The top layer of soil above the subsoil.

Transplant. A tree that has been grown for an additional year or more in a bed, before planting into the field.

Transpiration. The losing of water through the leaves and bark of a plant.

Tubeling. A seedling that has been grown in a small container.

Viability. The ability of a seed to germinate.

Water soluble. Relates to fertilizers and other materials that can be dissolved in water and applied to the plant with a sprinkler or sprayer.

Water stress. Condition in which a plant loses water faster than it absorbs it.

Weed. Any vegetation growing where it isn't wanted.

Weeding and thinning. The removal of unwanted trees and brush in a Christmas tree planting.

Wilding. Natural seedling of forest trees that is removed and transplanted for growing in plantations.

Wilts. Various plant diseases that cut off moisture to plants, sometimes causing death.

Windbreak. Trees and plants grown close together to provide shelter from the prevailing winds.

Suggested Further Reading

Although some of the following books are out of print, you may be able to find them at book sales or at your public library. Ask your county forester and Extension Service (or Agriculture Canada) for bulletins and other information.

Books

Audubon Society Staff and Elbert L. Little. *Audubon Field Guide to North American Trees*. Eastern and Western editions. Knopf, 1980.

Benyus, Janine M., editor. *Christmas Tree Pest Manual*. United States Department of Agriculture, 1983.

Burch, Monte. *Building Small Barns, Sheds, and Shelters*. Garden Way Publishing, 1983.

Chapman, Arthur G., and Robert D. Wray. *Christmas Trees for Pleasure and Profit*. Rutgers University Press, 1985.

Hill, Lewis. *Pruning Simplified*. Garden Way Publishing, 1986.

Wyman, Donald. *Trees for American Gardens*. Macmillan, 1965.

Yepsen, Roger B., Jr., editor. *The Encyclopedia of Natural Insect and Disease Control*. Rodale Press, 1984.

Magazines

American Christmas Tree Journal
611 East Wells Street
Milwaukee, WI 53202-3891

American Tree Farmer
American Forest Council
1619 Mass. Ave., NW
Washington, DC 20036

Journal of Forestry
Society of American Foresters
5400 Grosvenor Lane
Bethesda, MD 20814

Index

Spade, *56*
Species selection, 14, 25-38
  varieties and subspecies, 26-28
Spittlebugs, 94-95
Sprayers and spraying, 87-88
Spruces, 35-38. *See also* Black Hills
  spruce; Colorado blue spruce;
  Engelmann spruce; Norway
  spruce; Red spruce; Serbian
  spruce; White spruce
Stump culture, 100, *101,* 102
Sulfur as soil supplement, 72
Swiss needlecast. *See* Needlecasts

T

Tax considerations, 132-33
Theft, 97-98
Thrips, 95
Thuja occidentalis. See American
  arborvitae; Eastern white ce-
  dar
Time required for Christmas tree
  growing, 19-23
Tip greens, 104
Transplant bed: for seedlings, *48,* 49
Transplanting, *56. See also* Seedlings
  and transplants
Tubelings, *40,* 40-41

U

United States Department of Agriculture grading system, 118-19
Uredinopsis rust. See Rusts

V

Virginia pine, *31,* 31

W

Weather, 83-84
Webworms, 95
Weed control, 79-82
Weevils, 95-96
White fir. *See* Concolor white fir
White pine, *31,* 31
White spruce, 15, *36,* 37-38, *38*
Wholesale buyers, 120-21
Wild native tree plantations, 63-66
  nitrogen shortages in, 65-66
  pruning and shearing, 66
Wild tree farms, 6-7
Wildings, 41-42
Winter-burn, 84
Wreaths, 106-13
  decorating, 110-12
  handmade vs. machine-made, 106-
    7, *109*
  marketing, 117-26
  storing, 110
  supplies needed for, 108-10
  workplace for making, 107-8, *111*

*Page numbers in italics indicate that
an illustration appears on that page.*